Material Modeling in Finite Element Analysis

Finite element analysis has been widely applied in mechanical, civil, and biomedical designs. This new edition provides the readers with comprehensive views of various material models through practical examples, which will help them better understand various materials and build appropriate material models in finite element analysis. *Material Modeling in Finite Element Analysis*, Second Edition, consists of four main parts: (1) metals, (2) polymers, (3) soils, and (4) modern materials. Each part starts with the structure and function of different materials and then follows the corresponding material models and the temperature and time effects on the material models. The final part focuses on user subroutines such as UserMat and UserHyper. This book presents some specific problems including the metal-forming process, combustion room, Mullins effect of rubber tires, viscoelasticity of liver soft tissues, small punch test, tunnel excavation, slope stability, concrete slump test, orthodontic wire, and piezoelectric microaccelerometer. All modeling files are provided in the appendices of this book. This book would be helpful for graduate students and researchers in the mechanical, civil, and biomedical fields who conduct finite element analysis. This book provides all readers with a comprehensive understanding of modeling various materials.

Zhaochun Yang received his PhD in Mechanical Engineering from the University of Pittsburgh in 2004. Since 2005, he has worked for big companies and national labs such as Nation Energy Technology Laboratory. He has been in the field of finite element analysis for over 20 years and has gained much experience, especially in material modeling. Up to now, he has published 12 journal papers and 3 books.

Material Modeling in Finite Element Analysis

Second Edition

Zhaochun Yang

CRC Press
Taylor & Francis Group
Boca Raton London New York

CRC Press is an imprint of the
Taylor & Francis Group, an **informa** business

Designed cover image © 123RF

Second edition published 2024
by CRC Press
2385 NW Executive Center Drive, Suite 320, Boca Raton FL 33431

and by CRC Press
4 Park Square, Milton Park, Abingdon, Oxon, OX14 4RN

CRC Press is an imprint of Taylor & Francis Group, LLC

First edition published by Taylor and Francis 2019

ISBN: 978-1-032-56602-3 (hbk)
ISBN: 978-1-032-56603-0 (pbk)
ISBN: 978-1-003-43631-7 (ebk)

DOI: 10.1201/9781003436317

Typeset in Times
by codeMantra

Contents

PART II Polymers

PART III Soil

PART IV Modern Materials

PART V Retrospective

Preface

Three years ago, inspired by the success of my first book, *Finite Element Analysis for Biomedical Engineering Applications*, I finished my second book, *Material Modeling in Finite Element Analysis,* to summarize my experience in material modeling. Soon after the second book was published, I questioned my decision to write and publish it because material modeling is a huge field and I only know a tiny part of it. Therefore, I devoted the past 3 years to studying material modeling to develop a deeper understanding of it. I also realized that more and more parts of the second book should be rewritten. Thus, in the past 2 months, I collected all materials and rewrote a new version of my book, *Material Modeling in Finite Element Analysis.* The major modifications include:

1. Reorganization of Part I in order of the material models, effect of time and temperature on material properties, and user subroutine;
2. Rewriting of user subroutines, UserMat and UserHyper, because they are the most important user material subroutines;
3. Clarification of the differences between the materials, such as Drucker-Prager vs Mohr-Coulomb, BISO vs BKIN, and creep vs viscoplasticity; and
4. Deletion of some unnecessary examples and addition of more practical problems, including the strain-hardening model, Veronda-Westman model, small punch test, and breast tumor model. Some typical problems, such as soil excavation, concrete slump test, and one-dimensional Terzaghi's problem, to help understand the geomechanics models are also included.

During the writing and publication of the new version of the book, I received help from many of my friends: Frank Marx, Dr. J.S. Lin, Dr. Jobie Gerken, Dr. Adi Adumitroaie, Dr. Rachmadian Wulandana, Dr. Zhi-Hong Mao, Dr. Krystyna Gielo-Perczak, Dr. Rika Carlsen, and Ronna Edelstein. Without their help, I could not have completed my revision process. I also greatly appreciate editor Marc Gutierrez, editorial assistant Kianna Delly, and other staffs of CRC Press for their assistance in publishing this book. Finally, I welcome any feedback for improving this book.

1 Introduction

We live in a modern world of buildings, cars, airplanes, and medical advances, including coronary angioplasty. Buildings require structural analysis from civil engineers, while the manufacture of cars and airplanes depends on strong mechanical design. Biomedical engineering plays a significant role in coronary surgeries. All these fields demand an understanding of the stress state of the structures. With the development of computer technology, finite element analysis has been widely used in these fields. Specifically, since the 1970s, many commercial finite element software companies have evolved: ANSYS, ABAQUS, COMSOL, ADINA, LS-DYNA, and MARC. Among these, ANSYS emerges as the leader.

One of the key processes in finite element analysis is building material models. The structural analysis of a building is associated with soils; the manufacture of a car involves metals; the airplane has many components made of composites to ensure both high strength and low weight; and the stent in the coronary angioplasty is made of shape memory alloys. When we create finite element models to conduct stress analysis for their designs, we must adopt a variety of material models corresponding to different materials, because under the same geometry and loadings and boundary conditions, different materials may have unique behaviors. For example, it is hard to pull a steel rod, but the same force can easily deform a rubber rod. If we pull the metal rod to plasticity, the plastic strains remain after the force is gone. However, when the metal rod is made of shape memory alloys, the remaining transformation strains can be removed with the rise of temperature. Unlike these materials, the soil has little resistance to tension. The purpose of this book is to acknowledge that different materials have unique features and to present readers with a comprehensive view of various material models. By providing practical examples, this book will enable readers to better understand the material models, how to select appropriate material models, and how to define the correct solution set in the finite element analysis of complicated engineering problems. As discussed above, ANSYS is the leading software in the field of finite element analysis. Thus, the examples in this book are provided in the form of ANSYS input files, which makes it convenient for the readers to learn and practice these examples.

This book consists of four main parts. After the introductory chapter, the first part focuses on metals and alloys. Chapter 2 introduces the structure of alloys and their mechanical features. Some plastic models such as the isotropic hardening, kinematic hardening, and Chaboche model, along with ratcheting simulation, are presented in Chapter 3. Chapter 4 discusses the material properties, including viscoplasticity and creep, as a function of time. As the material properties of metals are temperature-dependent, Chapter 5 examines the influence of temperature on a combustion chamber. The last chapter in Part I shows how to develop UserMat and offers the strain-hardening model as an example.

DOI: 10.1201/9781003436317-1

The second part discusses polymers. After Chapter 7 depicts the structure and material properties of polymers, Chapter 8 presents some hyperelastic material models and curve-fitting of the models' material parameters. The viscoelasticity of elastomers and the shift functions are discussed in Chapter 9. Chapter 10 introduces eight chain-based viscoplasticity models, including application of the Bergstrom-Boyce model for simulation of the small punch test. The stress responses of elastomers always experience softening during the first few loading cycles. That is regarded as a damage accumulation in the material, which refers to the Mullins effect of elastomers; Chapter 11 focuses on this and includes an example of a rubber tire. UserHyper is available in ANSYS for customers to create their own hyperelastic models. Chapter 12 presents one example of UserHyper to reproduce the Veronda-Westman model.

Soils are the topic of the third part. Chapter 13 introduces the structure and various classifications of soils. Four major material models of soils—the Cam Clay model, Drucker-Prager model, Mohr-Coulomb model, and Jointed Rock model—are discussed in Chapters 14–17, respectively. These chapters also cover some typical and practical problems: excavation of the soil, a tower on the ground, a concrete slump test, soil-arch interaction, stability of a slope, and a tunnel excavation. Since soils are composed of rocks, water, and air, Chapter 18 explores the consolidation of soils and its application for simulation of the one-dimensional Terzaghi's problem and consolidation of rocks with three wells.

Part IV highlights modern materials. One widely used modern material is composite, which was developed in the 1950s due to the airplane design requiring high strength and low weight. Structure and material properties of composites are presented in the first part of Chapter 19, followed by an application on an adapter in flight-qualification testing and crack growth in the single-leg bending problem.

Chapter 20 examines functionally graded materials and their simulation in ANSYS using TBFIELD technology.

The unique features of shape memory alloys (SMAs), superelasticity, and shape memory effect are discussed in Chapter 21. The chapter includes two examples to demonstrate SMAs' applications: (1) orthodontic wire using the superelasticity feature and (2) a vacuum-tight shape memory flange using shape memory effect.

Chapter 22 reviews the structure and mechanical behavior of piezoelectric materials. This is followed by the simulation of a thin-film piezoelectric microaccelerometer using the piezoelectric material model in ANSYS.

Nanoscale materials refer to a group of substances with at least one dimension less than approximately 100 nm that attract more and more interest. The first part of Chapter 23 introduces nanoscale materials; in the second part, Young's modulus of nano-Fe particles is determined from the experimental data of a ball made of Fe particles using the optimization algorithm in ANSYS.

Chapter 24 reviews the features of metals/alloys, polymers, soils, and modern materials and discusses the relation between material properties and structures, as well as the relation between temperature and material properties. It also examines the solution control for various materials.

Part I

Metal

Metal plays a dominant role in modern industry, especially in the manufacturing of cars, airplanes, and ships. Its mechanical design requires a clear understanding of its stress and strain states. Therefore, Part I focuses on many metal material models implemented in ANSYS.

Chapter 2 introduces the structure and material properties of metal. Some plastic models, including the isotropic hardening and kinematic hardening models, are discussed in Chapter 3. Chapter 4 focuses on viscoplasticity and creep of the metal. The influence of temperature on the material properties is analyzed in Chapter 5. User subroutine and implementation of the strain-hardening model are covered in Chapter 6.

DOI: 10.1201/9781003436317-2

2 Structure and Material Properties of Metal

Material properties of metal are closely linked with its structure. Thus, this chapter first presents the structure of metal and then gives the material properties of metal.

2.1 STRUCTURE OF METAL

A metal is a material in which the atoms are joined together by metallic bonds (see Figure 2.1). The delocalized electrons and the strong interaction forces between the positive atom nuclei give the metal features such as good thermal and electrical conduction.

Assuming the material's atoms as a perfect crystal, the bond energy is expressed as [1]

$$U = \frac{-ACe^2}{r} + \frac{B}{r^n} \tag{2.1}$$

where A, B, C, and e are material constants.

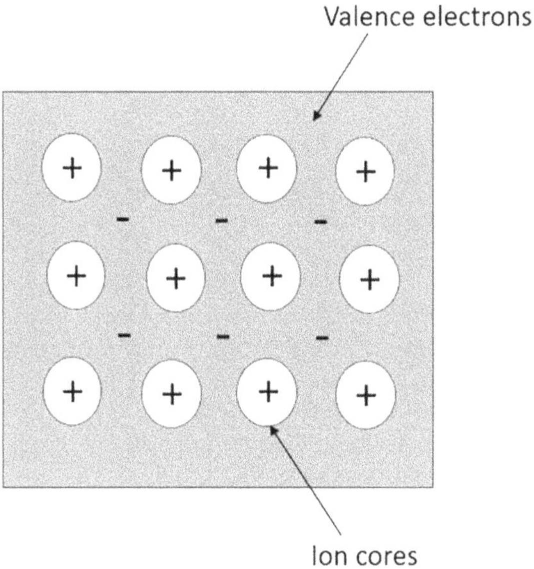

FIGURE 2.1 Atoms are held together by metallic bonds in a metal.

DOI: 10.1201/9781003436317-3

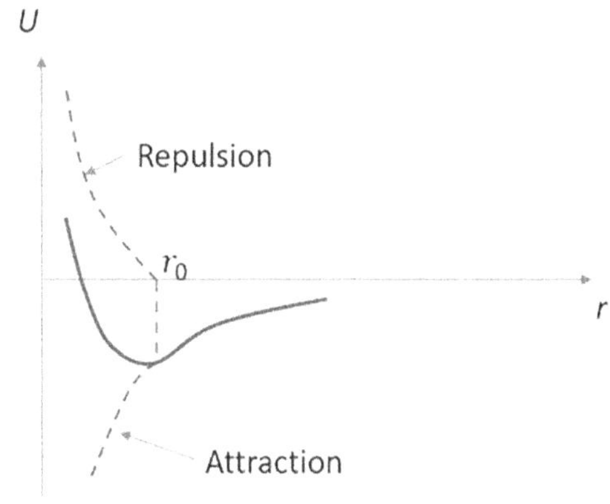

FIGURE 2.2 Bond energy function.

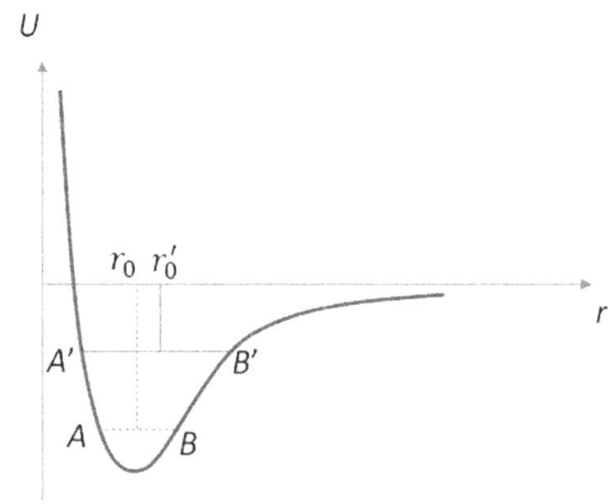

FIGURE 2.3 Anharmonicity of the bond energy function [1].

As shown in Figure 2.2, the curve is more flattened out at larger separation distances. When the distance between atoms is less than r_0, the forces between the atoms are repulsive. When the distance between atoms is greater than r_0, the forces become attractive.

Anharmonicity of the bond energy function explains why materials expand when the temperature increases (see Figure 2.3). As the internal energy rises due to the addition of heat, the system oscillates between the new positions marked *A*' and *B*' from the original positions *A* and *B*. Since the curve is not shaped like a sine curve, the new separation distance is longer than the previous one. Thus, the material has expanded, thereby causing the thermal strain.

2.2 ELASTICITY AND PLASTICITY OF METAL

When a force is applied to the metal, the layers of atoms start to roll over each other. If the force is released, the layers of atoms fall back to their original positions. Then, the metal is regarded as elastic (see Figure 2.4a).

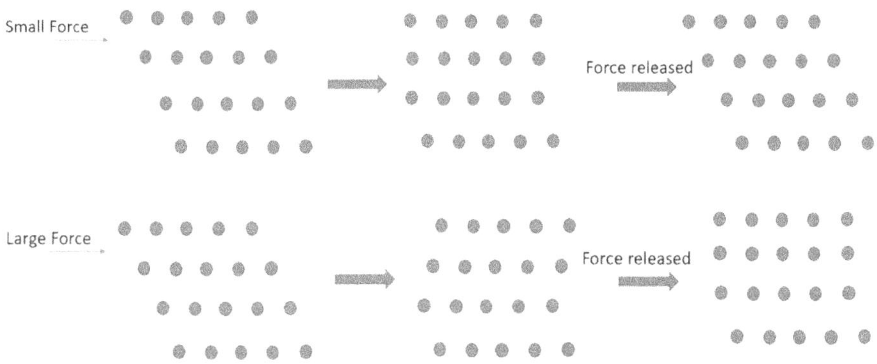

FIGURE 2.4 Illustration of a metal in elastic and plastic stages: (a) elastic stage and (b) plastic stage.

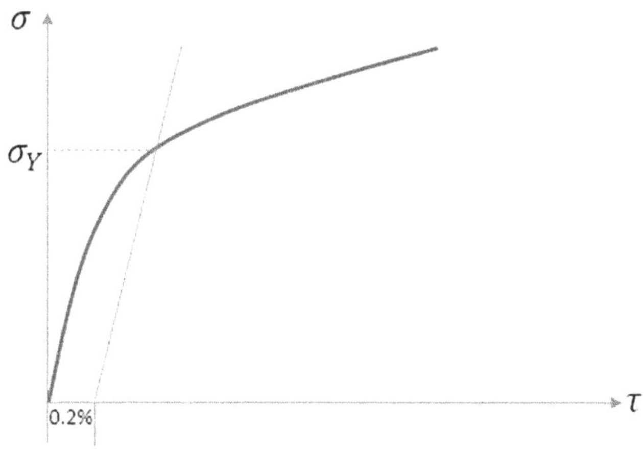

FIGURE 2.5 Stress-strain curve of the metal.

If the force is big enough that the atoms cannot fall back to their starting positions after the force is released, the metal is permanently changed (see Figure 2.4b).

Based on the above statements, the uniaxial tension tests of the metal show that when the stresses of the metal are below the yield stress σ_Y, the metal is elastic. When the stresses of the metal are higher than the yield stress σ_Y, the metal yields (see Figure 2.5). Yield driven by shearing stresses slides one plane along another (see Figure 2.6). The plastic deformation due to yield is a viscous flow process. Like flow in liquids, plastic flow has no volume change with Poisson's ratio $\upsilon = 0.5$.

Normally, if the material is in the elastic stage, it is not likely to fail. This is not true for brittle materials like ceramics because they fracture before they yield. However, for most of the structural materials, no damage occurs before yield. Thus, in the structural design, the materials required within the elastic stage use a safety factor. However, the plasticity of the metal has applications in industries such as metal forming. Therefore, some plastic models are introduced in Chapter 3.

FIGURE 2.6 Sliding is induced by shear stresses.

REFERENCE

1. Roylance, D., *Mechanical Properties of Materials*, MIT, Cambridge, 2008.

3 Some Plastic Material Models of Metals

After introducing the plasticity theory, Chapter 3 presents some isotropic hardening models and kinematic hardening models; it also includes a simulation of ratcheting in a notched bar.

3.1 INTRODUCTION OF PLASTICITY

The uniaxial tension test of the ductile material always has the typical stress-strain curve (see Figure 3.1), which shows the linear elasticity (*OA*), nonlinear elasticity (*AB*), the yield point *B*, the hardening stage from *B* to *D*, and the unloading stage *DE*. Figure 3.1 indicates that the total strain is composed of the elastic strain ε^{el} and plastic strain ε^{pl}:

$$\varepsilon = \varepsilon^{el} + \varepsilon^{pl} \tag{3.1}$$

The stress is determined by the elastic strain ε^{el}

$$\sigma = \mathbf{D}\varepsilon^{el} \tag{3.2}$$

where **D** is the stiffness matrix.

The general yield function is expressed as

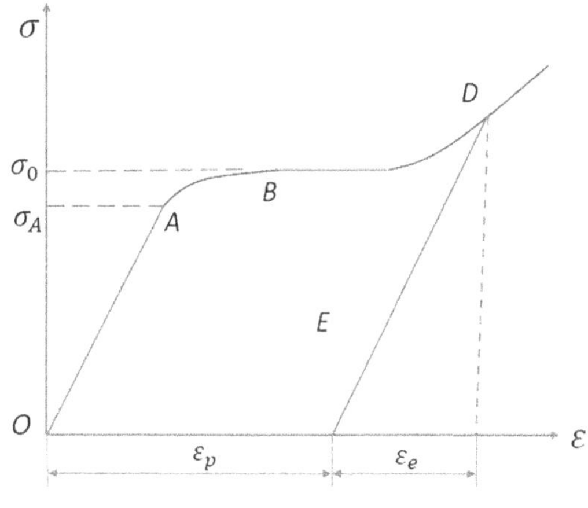

FIGURE 3.1 Stress-strain curve of the ductile material.

DOI: 10.1201/9781003436317-4

$$f(\sigma,\xi)=0 \tag{3.3}$$

where ξ refers to a set of internal variables.

When $f(\sigma,\xi)<0$, it is inside the yield surface and in an elastic state. No stresses exist outside the yield surface; instead, the stresses are either on or inside the yield surface.

Hardening occurs when an increase in stress leads to an increase in plastic strain. There are two types of hardening:

1. isotropic hardening of the yield surface (see Figure 3.2)
   The isotropic hardening changes the size of the yield surface but keeps the shape of the yield surface. It can model the behavior of the metals under monotonic loading. BISO and MISO in ANSYS belongs to isotropic hardening (see Figures 3.3 and 3.4), which can be defined by the following commands, respectively [1,2]:

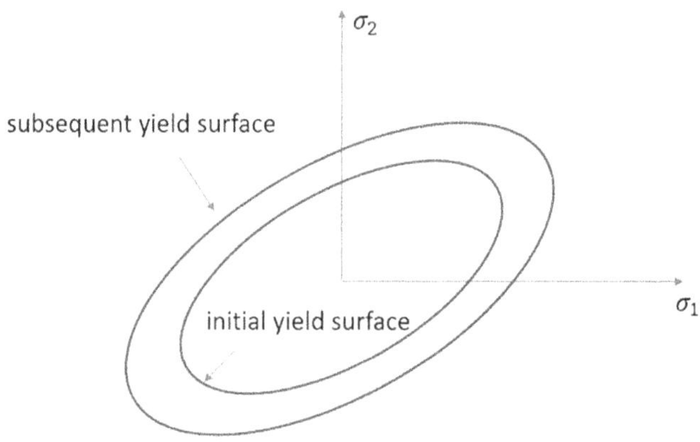

FIGURE 3.2 Isotropic hardening.

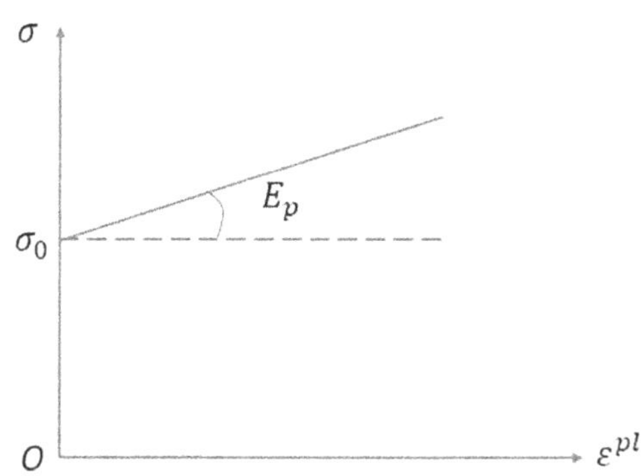

FIGURE 3.3 Plastic strain vs stress for bilinear isotropic hardening.

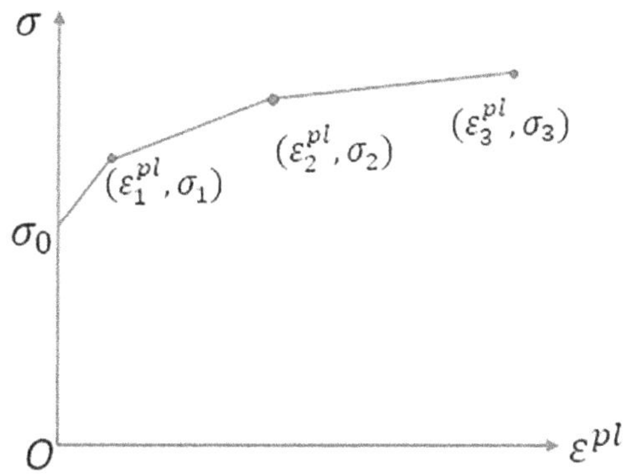

FIGURE 3.4 Plastic strain vs stress for multilinear isotropic hardening.

```
TB, Plastic, matid,BISO
TBDATA, 1, σ0, Ep
```

And

```
TB, Plastic, matid,MISO
TBPT, DEFI, 0, σ0
TBPT, DEFI, ε1pl, σ1
TBPT, DEFI, ε2pl, σ2
...
TBPT, DEFI, εipl, σi
```

2. kinematic hardening of the yield surface (see Figure 3.5)

Kinematic hardening changes the shape of the yield surface, such as BKIN and KINH in ANSYS. Its yield criteria become

$$f(\sigma - \alpha, \xi) = 0 \tag{3.4}$$

where α is the back stress that shifts the position of the yield surface in stress space during plastic deformation.

Their definition in ANSYS is like that of BISO and MISO, respectively.

Next, we conducted a uniaxial cyclic loading simulation to show the difference between isotropic hardening (BISO) and kinematic hardening (BKIN). The cyclic loading with time is illustrated in Figure 3.6, and the results (stress-strain curve) are presented in Figure 3.7; this clearly indicates that the BISO model increases yield surface, and the BKIN model only moves the yield surface center but does not change the size of the yield shape.

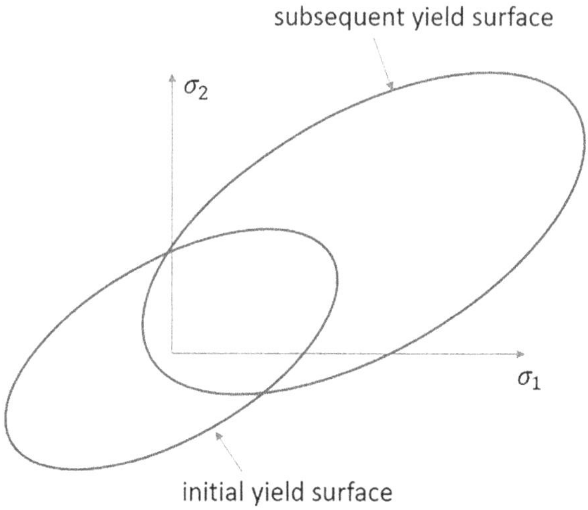

FIGURE 3.5 Kinematic hardening.

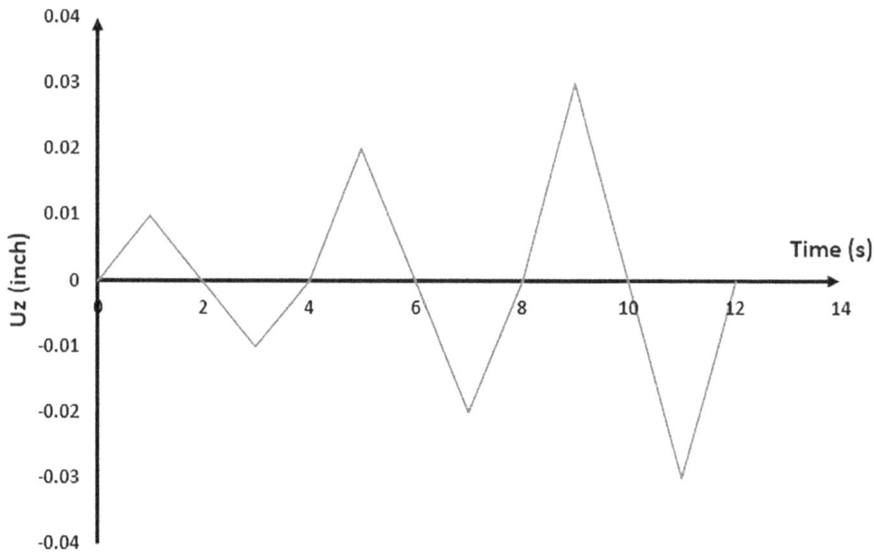

FIGURE 3.6 Displacement loading of the uniaxial test.

3.2 NONLINEAR KINEMATIC HARDENING

3.2.1 Introduction of the Chaboche Model

Under cyclic mechanical loading, the plastic deformation accumulates, and the stress-strain curve has a shift of the stress-strain hysteresis loop along the strain axis (see Figure 3.8), which is called ratcheting. Since the strain in ratcheting increases with the cyclic number of the loading, ratcheting may cause structure failure due

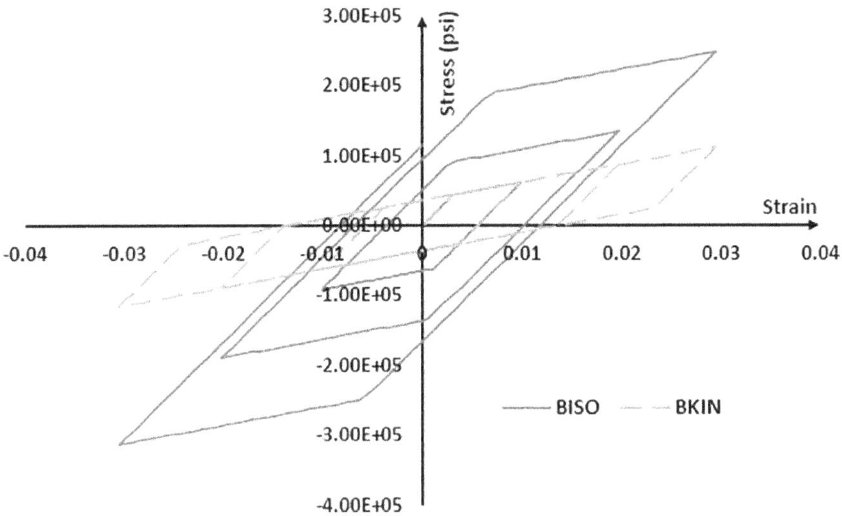

FIGURE 3.7 Stress-strain curves of the BISO and BKIN models under cyclic loading.

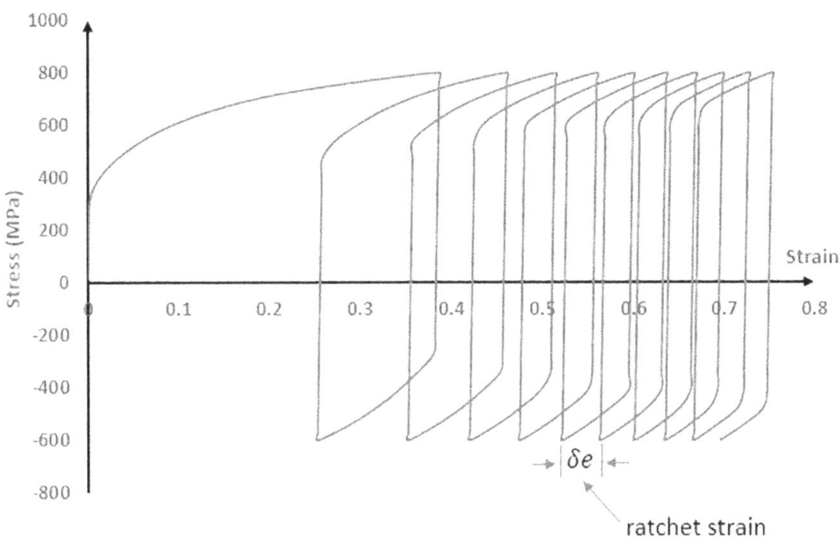

FIGURE 3.8 Ratcheting.

to structural instability; therefore, the study of ratcheting attracts more and more interest. The simulation of ratcheting demands that incremental plastic deformation terms be added to the general cyclic stress-strain equations. Thus, models like the multilinear kinematic hardening model, such as the BKIN and KINH models, cannot predict ratcheting. A few models, including the Chaboche model, are proposed to simulate ratcheting.

Chaboche and Rousselier [3] found that the hardening behavior of the steel material can be better approximated by the sum of various Frederick-Armstrong formulas. The back-stress tensor in the Chaboche model is expressed as

$$\alpha = \sum_{i=1}^{n} \alpha_i \tag{3.5}$$

where n is the number of superposed kinematic models.

The evolution of each back-stress model is specified as

$$\dot{\alpha}_i = \frac{2}{3} C_i \dot{\varepsilon}^{pl} - \gamma_i \dot{\bar{\varepsilon}}^{pl} \alpha_i + \frac{1}{C_i} \frac{dC_i}{d\theta} \theta \dot{\alpha}_i \tag{3.6}$$

where $\dot{\varepsilon}^{pl}$ – plastic strain rate,

$\dot{\bar{\varepsilon}}^{pl}$ – magnitude of the plastic strain rate, and

C_i and γ_i – user-input material parameters.

The Chaboche model is defined in ANSYS using the following commands:

```
TB, Chaboche, 1
TBDATA, 1, σ0
TBDATA,2, C1, γ1
TBDATA, 4, C2, γ2
...
TBDATA, 2n, Cn, γn
```

The Chaboche model can work with any of the available isotropic hardening models. It is very useful for simulating cyclic plastic behavior such as ratcheting.

3.2.2 Curve-Fitting of the Parameters of the Chaboche Model

The n-term Chaboche model has $2n$ parameters that need to be defined before specifying the Chaboche model for modeling. We may achieve it using the curve-fitting tool available in ANSYS. The experimental data come from a low cycle fatigue test [4] (see Figure 3.9). First, the Young's modulus was determined from the elastic stage. Then, the plastic strain was obtained by subtracting the elastic strain from the total strain. With the available data of the plastic strain versus the stress, the curve-fitting was conducted in ANSYS using the following commands:

```
TBFT,EADD,1,UNIA,AMBI.EXP
TBFT,FADD,1,PLAS,CHAB,3 ! THREE-TERM CHABOCHE MODEL

TBFT,SET,1,PLAS,CHAB,3,1,1E6 !INITIALIZE THE PARAMETERS, C1
```

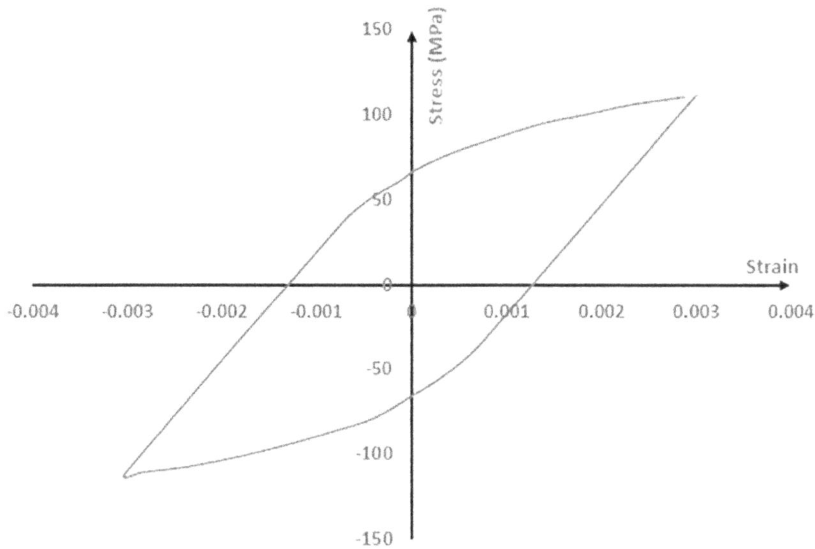

FIGURE 3.9 Experimental data for curve-fitting.

```
TBFT,SET,1,PLAS,CHAB,3,2,1E5 !INITIALIZE THE PARAMETERS, γ1
...

TBFT,SET,1,PLAS,CHAB,3,7,55 !INITIALIZE σ0 = HALF OF σmax
TBFT,SOLVE,1,PLAS,CHAB,3,0,1000
```

In the above initial value input, σ_0 was defined from half of σ_{max}. Other initial data only chose big values.

Curve-fitting results, which are plotted in Figure 3.10, show that the curve-fitting results effectively match the experimental data. Therefore, the Chaboche model is defined as

```
TB, CHABOCHE, 1, ,3
TBDATA, 1, 67.5
TBDATA, 2, 1.000E+06, 9.371E+04
TBDATA, 4, 1.077E+04, 9.994E+04
TBDATA, 6, 4.110E+04, 1.111E+03
```

3.2.3 Simulation of Ratcheting in a Notched Bar

In this example, a notched bar subjected to cycling loading was simulated to produce the plastic ratchetting in ANSYS190.

3.2.3.1 Finite Element Model

A 2-D axisymmetrical notched rod was modeled in ANSYS190 using Plane182 with keyopt (3)=1 (axisymmetrical) (see Figure 3.11). Also, the model was simplified using the symmetrical condition in the vertical direction.

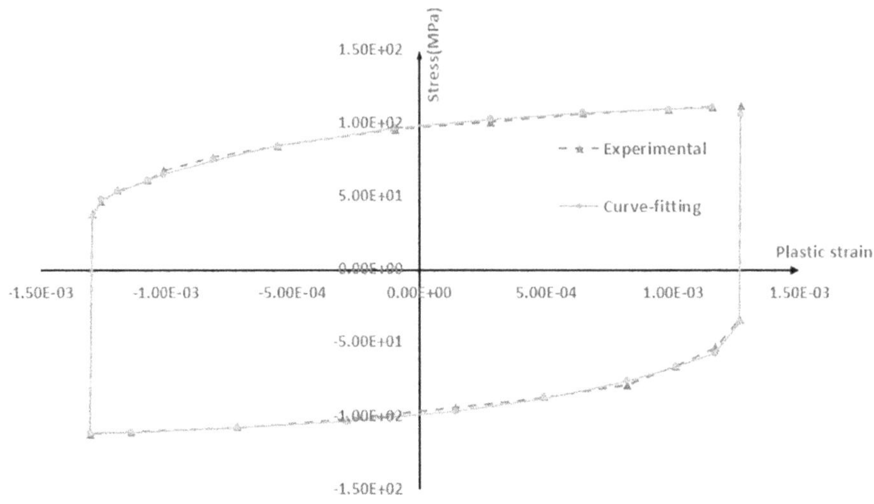

FIGURE 3.10 Curve-fitting results of the Chaboche model.

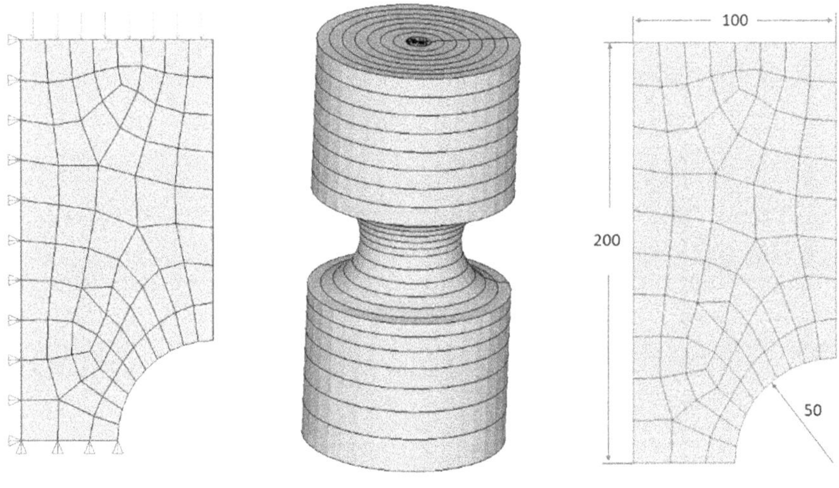

FIGURE 3.11 Finite element model of a notched rod (a) 2-D axisymmetrical, (b) 3-D, and (c) geometry (all dimensions in mm).

3.2.3.2 Material Properties

The rod was defined with the three-term Chaboche model plus the Voce isotropic hardening law (see Table 3.1). The material parameters of the Chaboche model were selected from the curve-fitting results of Section 3.2.

The material model was defined by the following commands:

```
POWER_N=0.1
SIGMA_Y=67.5
TB,NLISO,1,2,5
```

TABLE 3.1
Material Parameters of the Rod [5]

E (MPa)	ν	σ_0 (MPa)	C_1	γ_1	C_2	γ_2	C_3	γ_3	n
647,000	0.3	67.5	1e6	9.37e4	1e4	1e5	4.1e4	1.1e3	0.1

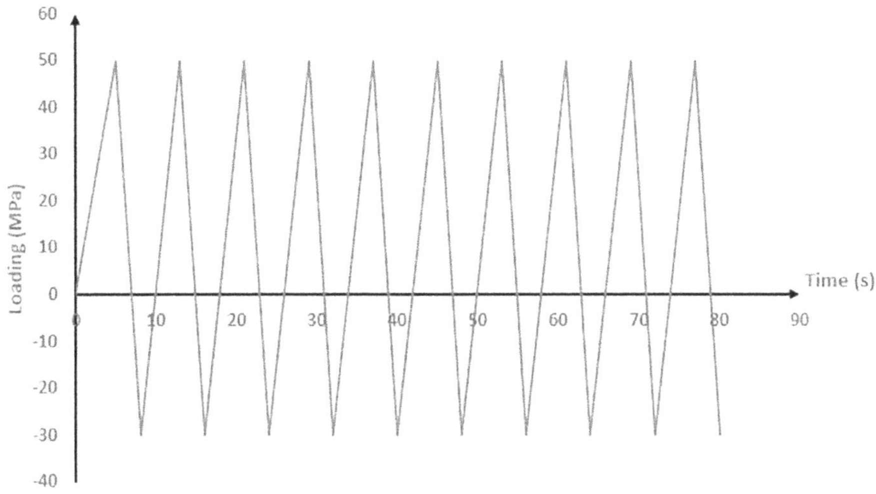

FIGURE 3.12 Loading history.

```
TBDATA,1,SIGMA_Y,POWER_N
TB,CHAB,1,3
TBDATA,1,SIGMA_Y,1E6,9.37E4,1E4,1E5
TBDATA,6,4.1E4,1.1E3
```

For comparison, the model was repeated with the bilinear kinematic hardening model, in which the tangent modulus E_T was chosen for 2,000 MPa.

3.2.3.3 Loadings and Boundary Conditions

The axisymmetrical condition and the symmetrical condition were applied on edges (see Figure 3.11a). The top edge was loaded with ten cycles of loading; each cycle had a maximum of 50 MPa in tension and a minimum of 30 MPa in compression (see Figure 3.12).

3.2.3.4 Results

The von-Mises plastic strain and stress of the rod at the end of the first cycle and the last cycle are plotted in Figures 3.13 and 3.14, respectively. The maximum values occur at the notched tip for both stresses and plastic strains. Comparing the results of the first cycle with those of the last cycle, the stresses do not change much because they are pressure-loaded, but the plastic strains change dramatically between the first and last cycles. Figures 3.15 and 3.16 depict the axial strain versus axial stress

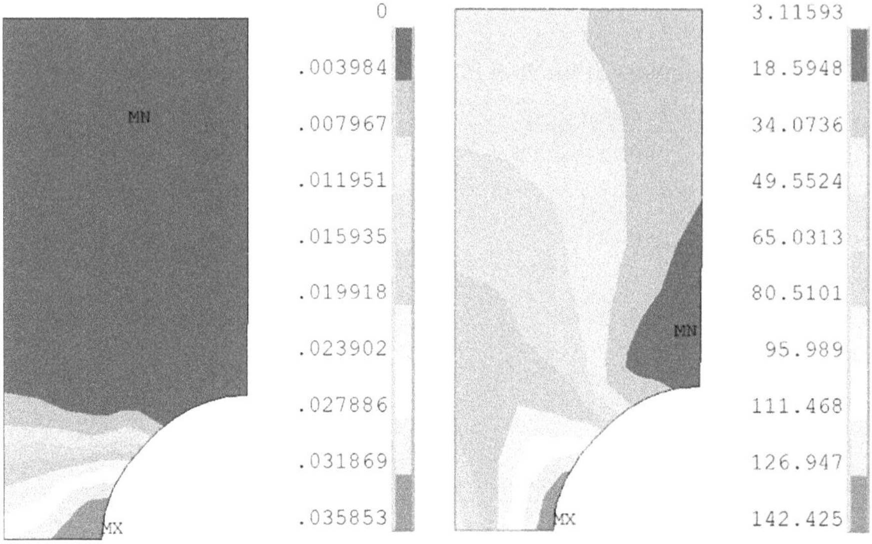

FIGURE 3.13 von-Mises plastic strains and stresses of the rod at the end of the first cycle: (a) von-Mises plastic strains and (b) von-Mises stresses (MPa).

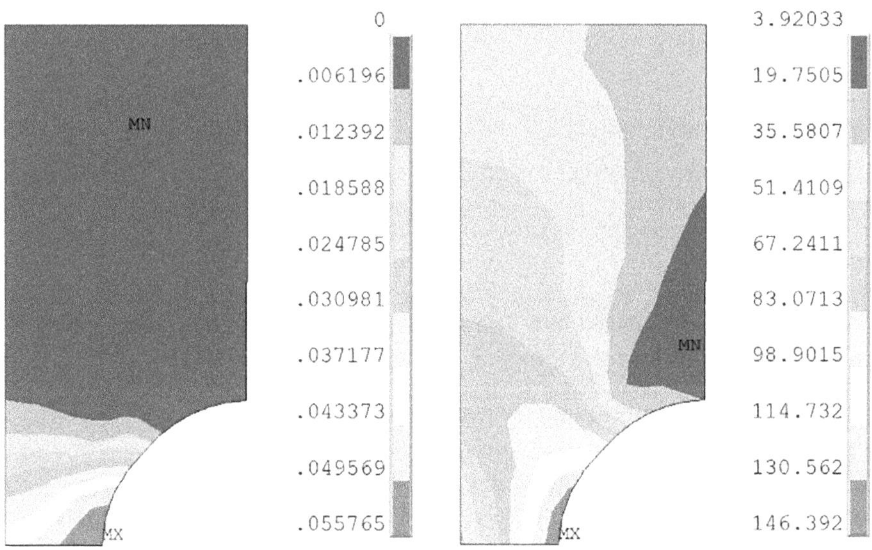

FIGURE 3.14 von-Mises plastic strains and stresses of the rod at the end of the last cycle: (a) von-Mises plastic strains and (b) von-Mises stresses (MPa).

with the Chaboche model and bilinear kinematic hardening model, respectively. Figure 3.15 clearly illustrates a shift of the stress-strain hysteresis loop along the strain axis, which is the ratchetting. On the other hand, no ratchetting appears in Figure 3.16. The difference between Figure 3.15 and Figure 3.16 confirms that the

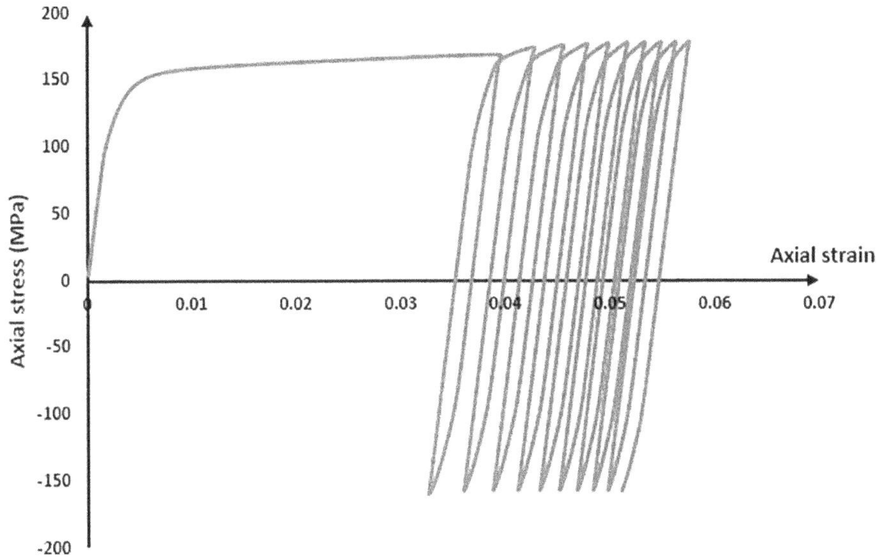

FIGURE 3.15 Axial strain vs axial stress.

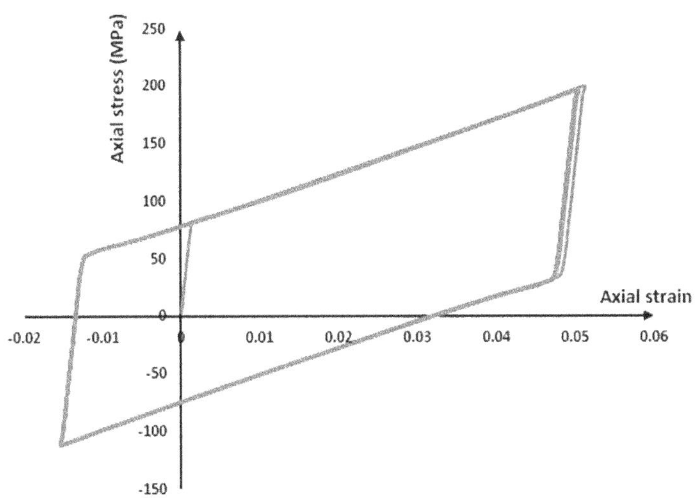

FIGURE 3.16 Axial strain vs axial stress with bilinear kinematic hardening.

ratchetting is modeled by the Chaboche model (nonlinear kinematic hardening), and the bilinear kinematic hardening model cannot produce the ratchetting of metal.

3.2.3.5 Summary

A notched bar with the Chaboche model under cycling loading was studied in ANSYS190. The relation between the axial strain and axial stress reveals the ratchetting due to the Chaboche model (nonlinear kinematic hardening).

REFERENCES

[1] ANSYS190 Help Documentation in the help page of product ANSYS190.

[2] Wang, J.D., and Ohno, N., Two equivalent forms of nonlinear kinematic hardening: application to nonisothermal plasticity, *International Journal of Plasticity*, Vol. 7, 1991, pp. 637–650.

[3] Chaboche, J.L., and Rousselier, G., On the plastic and viscoplastic constitutive equations, Parts I and II, *Journal of Pressure Vessel and Piping*, Vol. 105, 1983, pp. 453–164.

[4] Seisenbacher, B., Winter, G., and Grun, F., Improved approach to determine the material parameters for a combined hardening model, *Material Sciences and Applications*, Vol. 9, 2018, pp. 357–367.

[5] Kang, G., Finite element implementation of advanced constitutive model emphasizing on ratchetting, *18th International Conference on Structural Mechanics in Reactor Technology (SMiRT 18)*, Beijing, China, August 7–12, 2005.

4 Material Properties as Function of Time

Viscoplasticity and creep are material properties as functions of time. After discussing viscoplasticity with one example of metal forming, Chapter 4 then concentrates on creep and the simulation of creep of a bolt under pretension.

4.1 VISCOPLASTICITY

In the plastic deformation process, additional strain requires an increase in stress, which is known as Strain/Work hardening [1]. However, the hardening curves of the viscoplastic materials do not noticeably differ from those of rate-independent plastic materials. Three essential differences can be concluded from Figure 4.1:

- At the same strain, a higher rate of strain results in a higher stress.
- A strain rate change in the test causes an immediate change of the stress–strain curve.
- A plastic yield limit is not applicable.

4.1.1 Different Viscoplastic Models

ANSYS provides three viscoplastic models, including Perzyna, Peirce, and Exponential Visco-Hardening (EVH) [2].

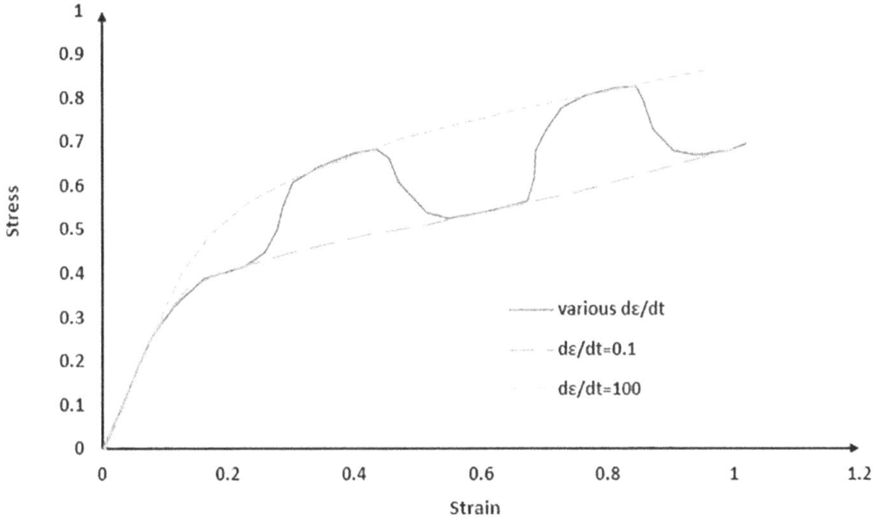

FIGURE 4.1 Stress–strain curve of a viscoplastic material at different strain rates [1].

DOI: 10.1201/9781003436317-5

Perzyna model:

$$\dot{\hat{\varepsilon}}_{pl} = \sum_{i=1}^{n} \gamma_i \left(\frac{\sigma}{\sigma_0} - 1 \right)^{\frac{1}{m_i}} \tag{4.1}$$

where n – the number of terms,

σ – the effective stress,
$\dot{\hat{\varepsilon}}_{pl}$ – the equivalent plastic strain rate,
m_i – the strain rate hardening parameters,
γ_i – the material viscosity parameters, and
σ_0 – the static yield stress.

Perzyna is defined in ANSYS by TB,RATE,,,,PERZYNA to input parameters m_i and γ_i.

Peirce model:

$$\dot{\hat{\varepsilon}}_{pl} = \sum_{i=1}^{n} \gamma_i \left(\left(\frac{\sigma}{\sigma_0} \right)^{\frac{1}{m_i}} - 1 \right) \tag{4.2}$$

Peirce is specified in ANSYS by TB,RATE,,,,PEIRCE to input parameters m_i and γ_i.

EVH model:

$$\dot{\hat{\varepsilon}}_{pl} = \sum_{i=1}^{n} \gamma_i \left(\frac{\sigma - \dot{\sigma}_0}{K_i} \right)^{\frac{1}{m_i}} \tag{4.3}$$

where K_i – the material viscosity parameters,

$\dot{\sigma}_0$ – the static yield stress that is defined by,

$$\dot{\sigma}_0 = K_0 + R_0 \hat{\varepsilon}_{pl} + R_\infty \left(1 - e^{-b\hat{\varepsilon}_{pl}} \right) \tag{4.4}$$

where K_0 – the initial yield stress,

R_0 – the linear hardening coefficient,
R_∞ – the exponential hardening coefficient, and
b – the saturation rate.

EVH is determined in ANSYS by TB, RATE,,,,EVH to input parameters K_0, R_0, R_∞, b, and K_i and m_i in order.

4.1.2 Simulation of Forming a Sheet

Metal forming refers to the metalworking process of metal parts through mechanical deformation; although the shape of the workpiece changes, its mass remains the same [3]. A tool called a die applies stresses over the yield strength of the metal. Thus, the metal takes a shape determined by the shape of the dies (see Figure 4.2). This section simulated the forming of a sheet between two dies in ANSYS190 with the EVH model.

4.1.2.1 Finite Element Model

A circle sheet was deformed between two dies. The right die was fixed, but the left one was moved down to make the sheet deformed. The whole model was simplified as a 2-D axisymmetrical model, and the sheet was meshed with Plane182 with keyopt(3)=1 (axisymmetrical) (see Figure 4.3).

4.1.2.2 Material Properties

The sheet was modeled with the first-term Chaboche model plus the Exponential Visco-Hardening (EVH) option (see Table 4.1). The Chaboche model was defined in ANSYS by the following commands:

```
TB,CHAB,1,,1                    ! DEFINE CHABOCHE MATERIAL
DATA
TBDATA,1,1.53e2,62511,1.1/311*62511
```

The EVH option was specified in ANSYS using the following commands:

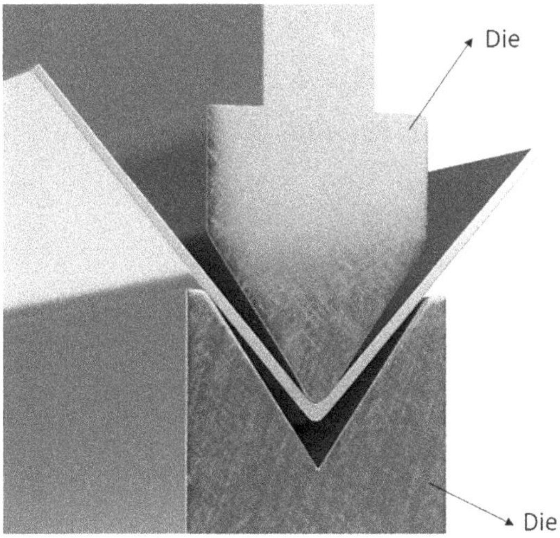

FIGURE 4.2 Forming of a sheet. (Eugene Zabugin © 123rf.com.)

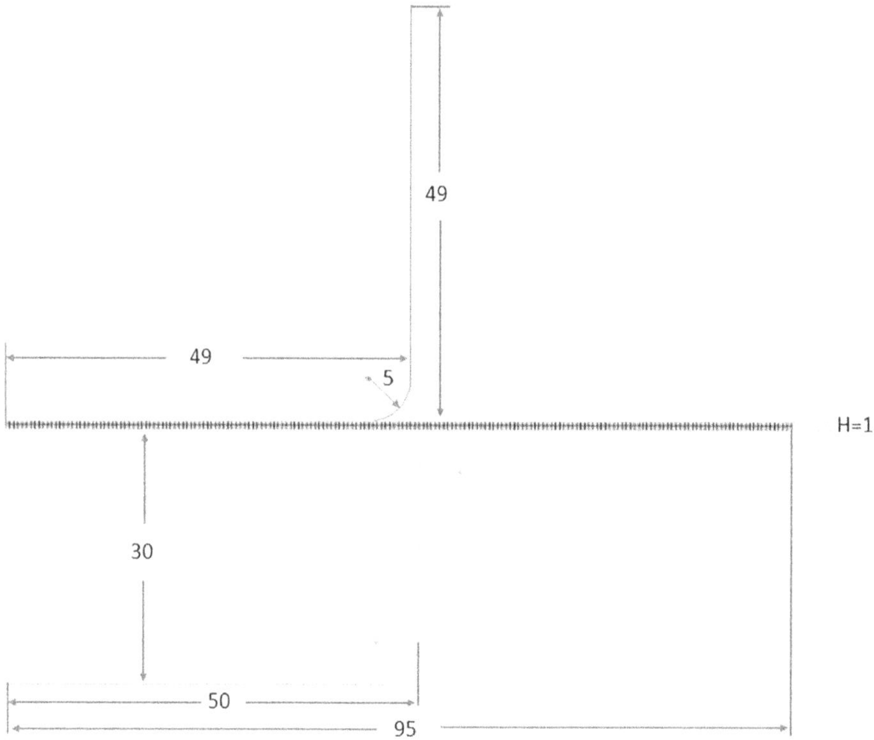

FIGURE 4.3 Finite element model of a sheet between two dies (all dimensions in mm).

TABLE 4.1
Material Parameters of the Sheet [4]

E (MPa)	v	σ_0 (MPa)	C_I	γ_I	R_0	R_∞	b	m_I	K_I
149,650	0.33	153	223	50	0	−153	317	1/7.7	1,150

```
TB, RATE, 1,,6,EVH                    ! DEFINE RATE DEPENDENT
MATERIAL DATA
TBDATA,1,1.53e2,0,-1.53e2,317,1/7.7,1150
```

4.1.2.3 Contact Definition

The contact between the sheet and the dies was assumed to be a standard contact. The dies were regarded as rigid.

4.1.2.4 Loadings and Boundary Conditions

The right die was fixed with all degrees of freedom, and the left one was moved down 30 mm (see Figure 4.4).

FIGURE 4.4 Boundary conditions.

4.1.2.5 Solution Setting

In the solution setting, RATE should be turned on for the EVH option. Furthermore, the command NLGEOM ON and the maximum substeps of 10,000 were defined because of the large deformation in the forming process. The entire forming process was assumed to be quasi-static without dynamic effects and finished in 1 second.

4.1.2.6 Results

The deformation of the sheet, including its 3D deformation, is plotted in Figure 4.5. von-Mises stresses and plastic strains of the sheet at the last step are presented in Figures 4.6 and 4.7, respectively. The maximum stress and plastic strain are localized around the top corner, with peak values of 1,470 MPa and 0.412, respectively. To show the effect of the EVH model, Figure 4.8 illustrates the time history of the reaction force in the vertical direction with/without the EVH model, which indicates that the reaction force with EVH is much greater than that without EVH. This also happens to the equivalent stress–equivalent strain curve of element 105 (see Figure 4.9).

4.1.2.7 Summary

The forming process of a sheet was modeled in ANSYS using the Chaboche model with the EVH option. The results show that the EVH model has a significant effect on the results.

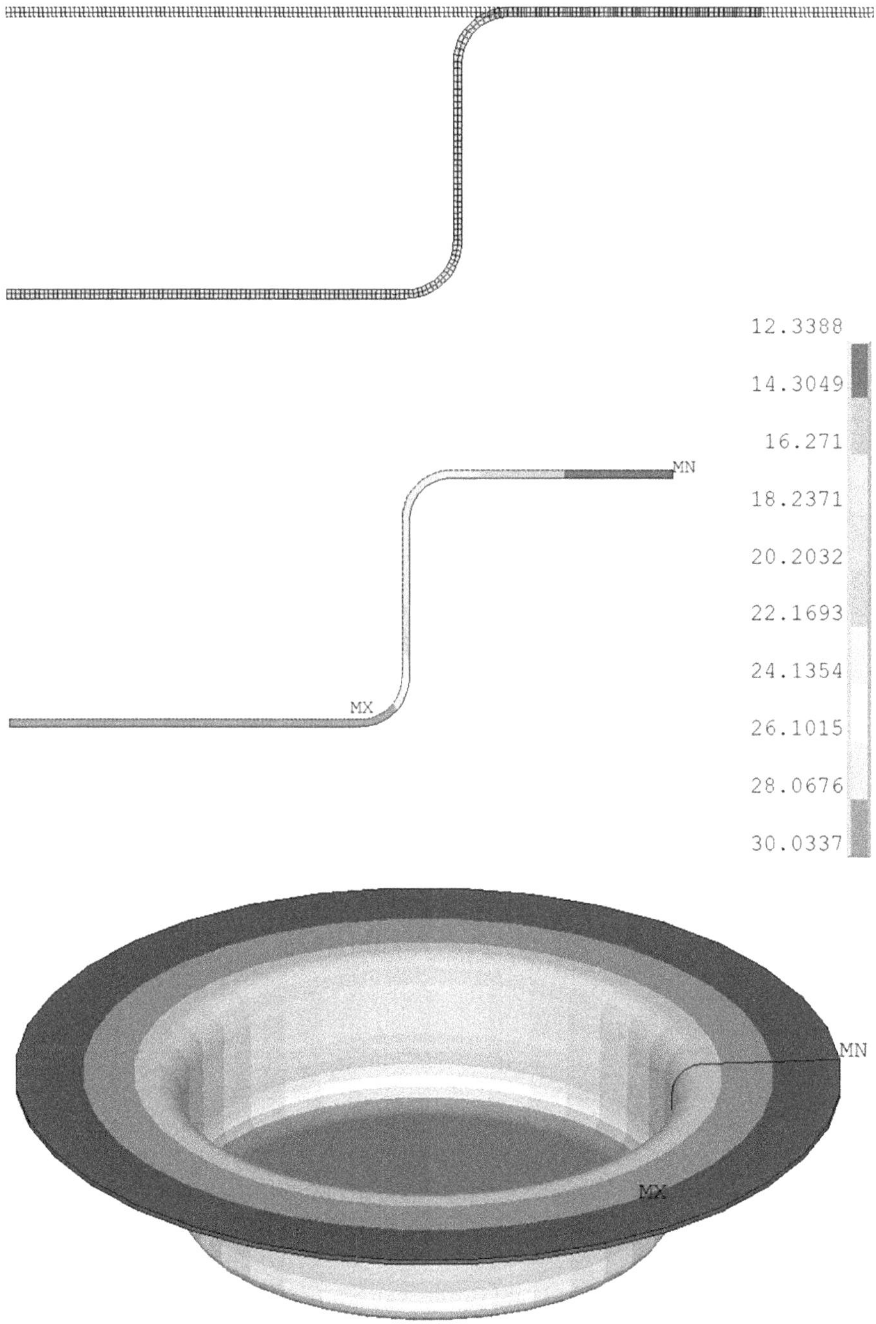

FIGURE 4.5 Final deformation of the sheet: (a) deformed shape vs original shape, (b) final displacement contour (2D axisymmetrical), and (c) final displacement contour (3D).

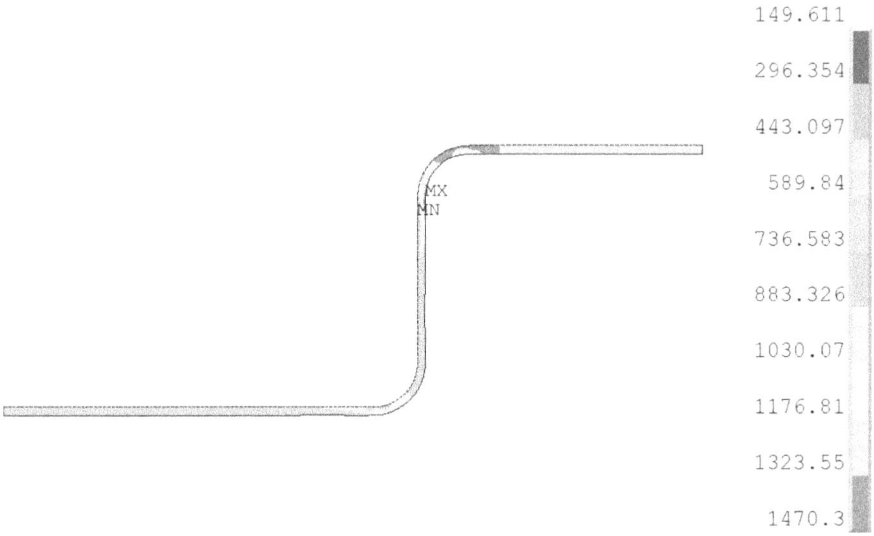

FIGURE 4.6 Final von-Mises stresses of the sheet (MPa).

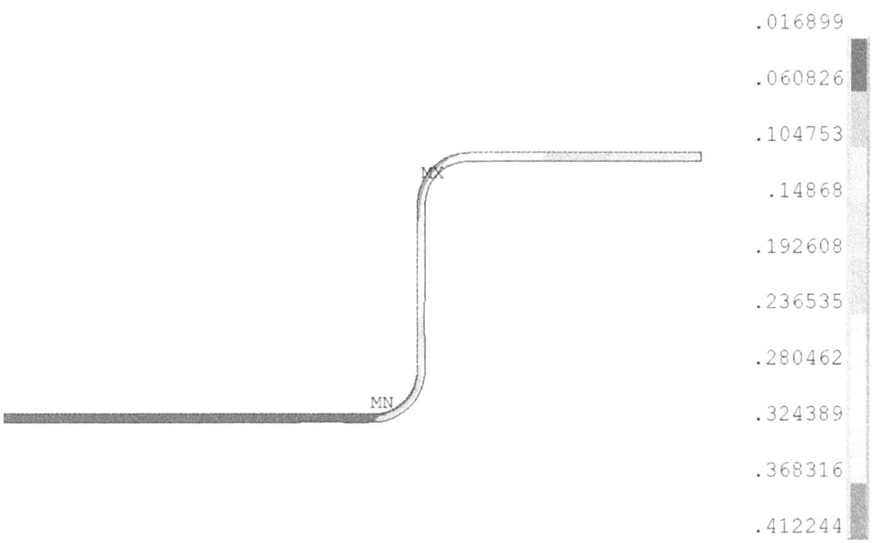

FIGURE 4.7 Final plastic strains of the sheet.

4.2 CREEP

Creep refers to the strains (stresses) of metals that, when under high loadings such as temperature, change with time in constant working conditions. Figure 4.10 illustrates the typical creep curve, composed of three stages [5]. The first stage is the primary creep stage, in which the creep starts at a rapid rate and slows gradually with time. The creep maintains a relatively uniform rate in the second creep stage; thus, the second creep stage is the steady stage. The third stage has an accelerating creep ratio that always

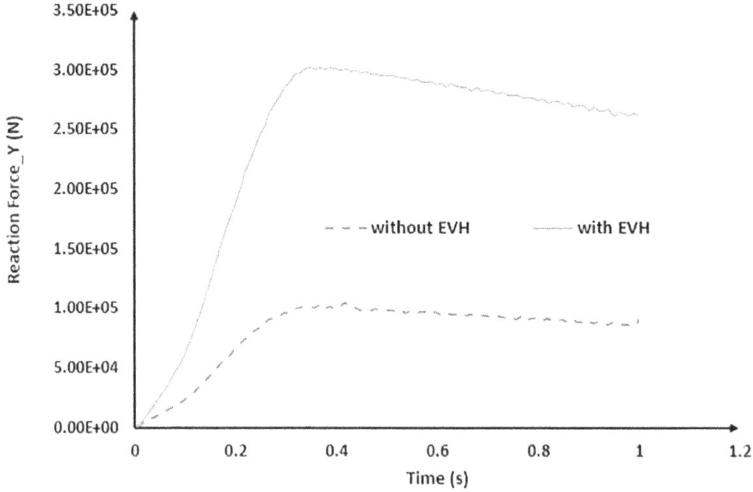

FIGURE 4.8 Time history of the reaction force in the Y direction.

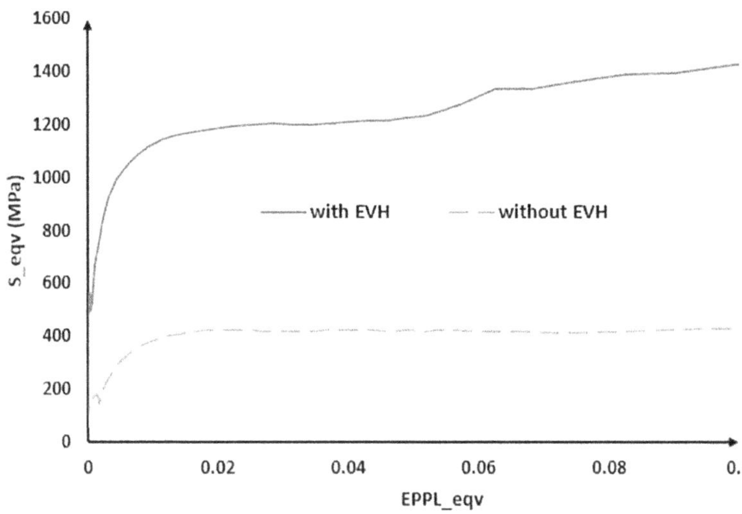

FIGURE 4.9 equivalent plastic strain–equivalent stress of element 105.

leads to material failure. Creep always occurs in metalworking, springs, soldered joints, and high-temperature materials. Two major cases exist relevant to creep:

1. Under constant loading, the creep strains of the structure increase with time. Thus, the deformations of the structure grow larger and larger (see Figure 4.11a).
2. Under initial stresses and fixed boundary conditions, stresses of the structure decrease with time. This process is called stress relaxation (see Figure 4.11b).

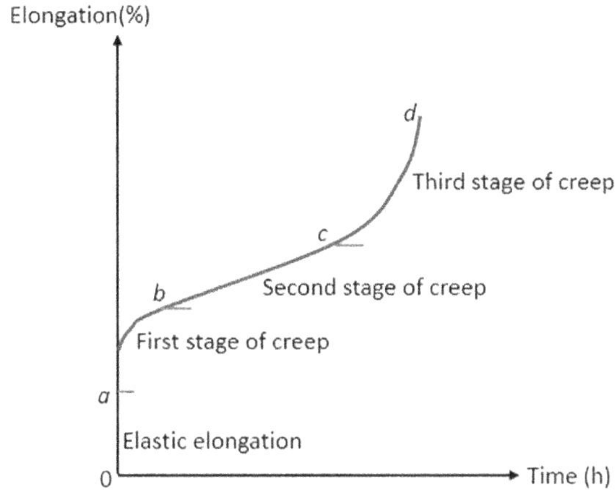

FIGURE 4.10 Typical creep curve.

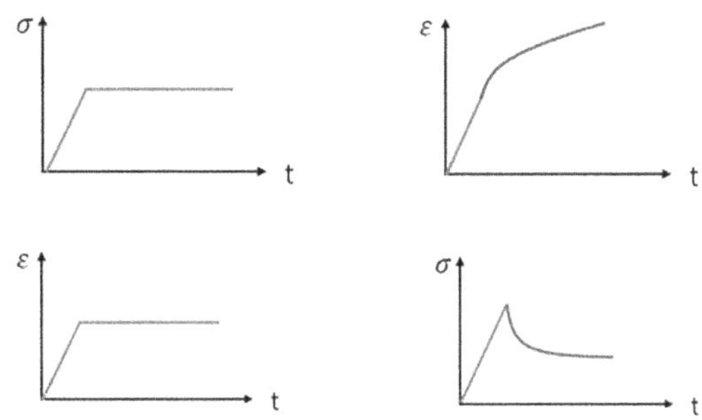

FIGURE 4.11 Two major creep behaviors: (a) creep and (b) stress relaxation.

4.2.1 Creep Constitutive Law

Generally, the creep equation is a function of stress, strain, time, and temperature,

$$\dot{\varepsilon}_{cr} = f_1(\sigma) f_2(\varepsilon) f_3(t) f_4(T) \tag{4.5}$$

ANSYS provides 13 implicit creep equations, including time hardening. The time hardening equation is expressed as [2]

$$\dot{\varepsilon}_{cr} = c_1 \sigma^{C2} t^{C3} e^{-\frac{C4}{T}} \tag{4.6}$$

The equation (4.6) indicates that the equation has four parameters – C_1, C_2, C_3 and C_4 – related to stress, time, and temperature, but no strain. C_2 defines the relation between the creep ratio and stress. C_2 is normally selected between 2 and 20. C_3 is the time factor.

The implicit creep analysis in ANSYS is robust, fast, accurate, and widely used in practice. Furthermore, the creep function can be used individually or combined with kinematic hardening plasticity. The ANSYS command to define the creep function is specified as follows [2]:

```
TB, CREEP, matid,TBOPT   ! TBOPT specifies the creep equation
TBDATA, 1, C1, C2, C3, C4
```

4.2.2 Simulation of Creep of a Bolt under Pretension

In Section 4.2.2, the creep of a bolt under pretension was studied in ANSYS190 using TB, CREEP command.

4.2.2.1 Finite Element Model

The model consists of two steel plates connected by a bolt. Because of the symmetrical condition, one half of the model was created and meshed by SOLID187 (see Figure 4.12).

4.2.2.2 Material Properties

Both the bolt and the steel plate were modeled as a MISO plastic model. In addition, the creep of the bolt was defined as time hardening using TB, CREEP command. The material models were defined using the following APDL commands:

```
! MATERIAL MODEL OF STEEL
MP,EX,1,14.665E6
MP,PRXY,1,0.30
TB,PLAS,1,1,2,MISO
TBPT,DEFI,0,29.33E3
TBPT,DEFI,1.59E-3,50E3

! MATERIAL MODEL OF BOLT
MP,EX,2,42E6
```

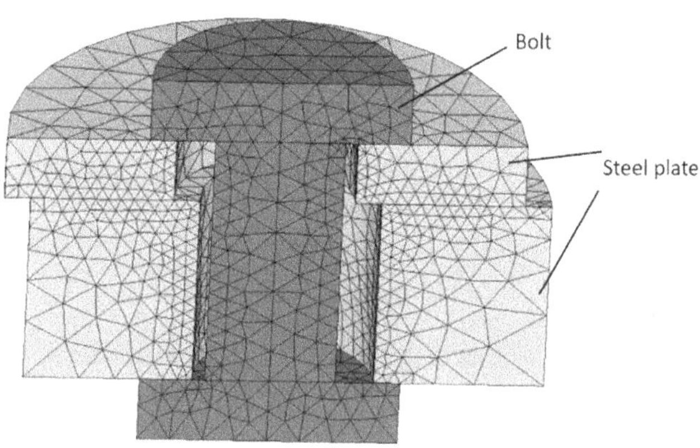

FIGURE 4.12 Finite element model of pretension between the plates and the bolt.

```
MP,PRXY,2,0.30
TB,PLAS,2,1,2,MISO
TBPT,DEFI,0,29.33E3*3
TBPT,DEFI,1.59E-3,50E3*3
TB,CREEP,2,1,4,2
TBDATA,1,0.54E-16,2.3,-0.6
```

4.2.2.3 Loadings and Boundary Conditions

In addition to the symmetrical condition, 0.01-inch pretension was applied on the bolt using the SLOAD command (see Figure 4.13).

4.2.2.4 Solution Setting

The solution has two steps. The first step only takes 1e-4 seconds to load the pretension. Because the creep calculation is always inaccurate during a very short time period, RATE was turned off in the first step to deactivate the creep. The second step with RATE turned on is creep time.

4.2.2.5 Results

Figures 4.14 and 4.15 illustrate the von-Mises stresses of the model at the end of the first step and the second step, respectively. The results indicate that stresses at the end of the last step are much less than those at the end of the first step, which is due to the creep of the bolt. This conclusion is confirmed by the time history of von-Mises stress of an element of the bolt (see Figure 4.16), which illustrates the case of stress relaxation mentioned in Figure 4.11b.

4.2.2.6 Discussion

The creep simulation is divided into two steps. The first step is the loading step without creep effects. Normally, the time of this first step is very short. The second step

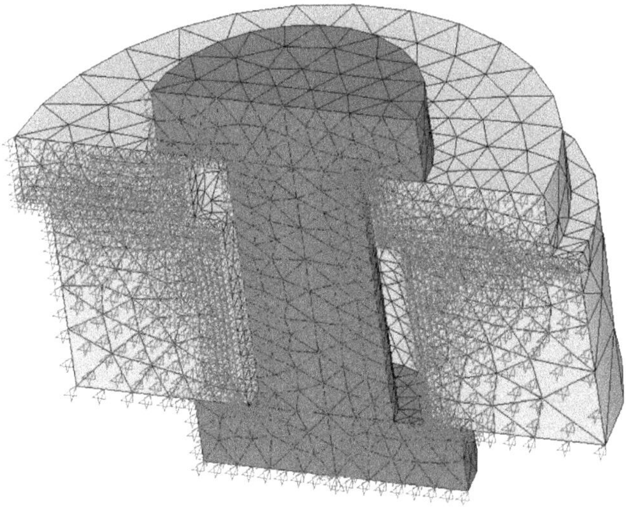

FIGURE 4.13 Boundary conditions.

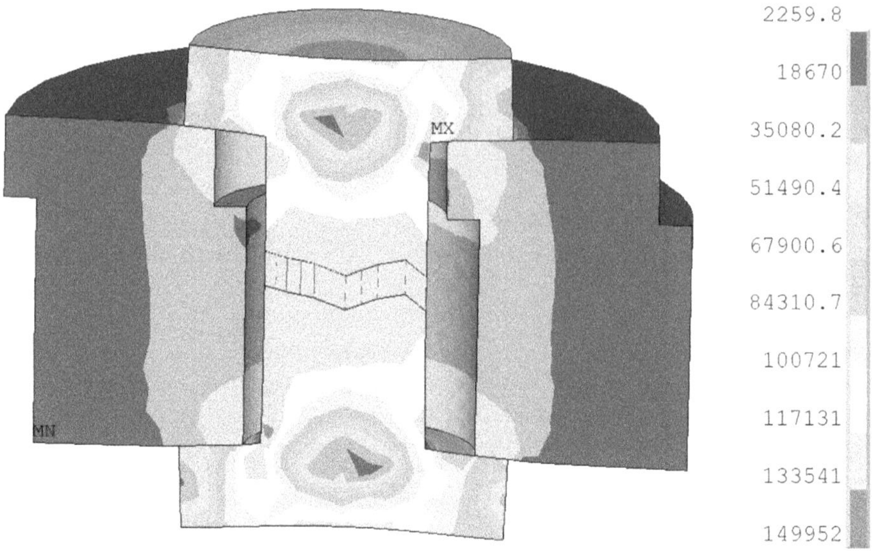

FIGURE 4.14 von-Mises stresses of the model after pretension (1st loading step) (Psi).

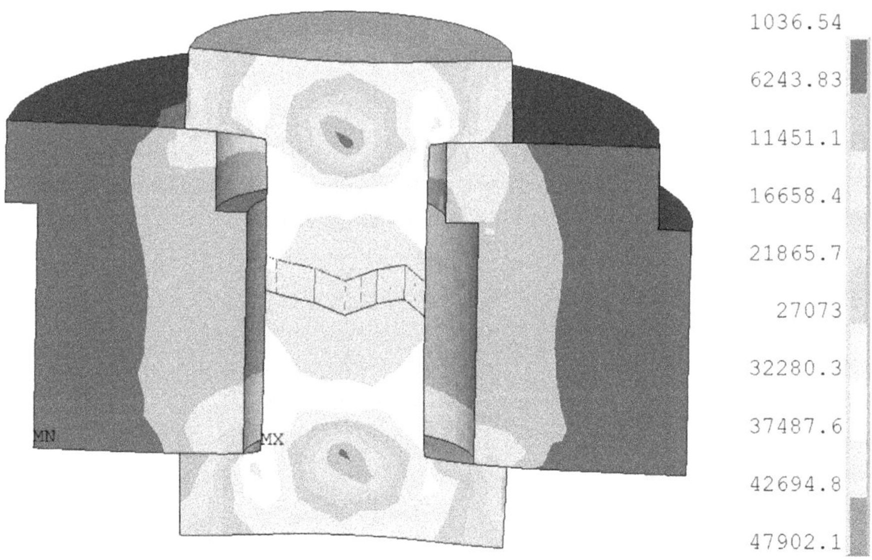

FIGURE 4.15 von-Mises stresses of the model after 17.36 days (Psi).

is the creep step with RATE turned on. The method of dividing the creep calculation into a loading step and a creep step is widely used in the creep calculation because it has been proven that a short time period may cause the calculation of creep strain to be inaccurate.

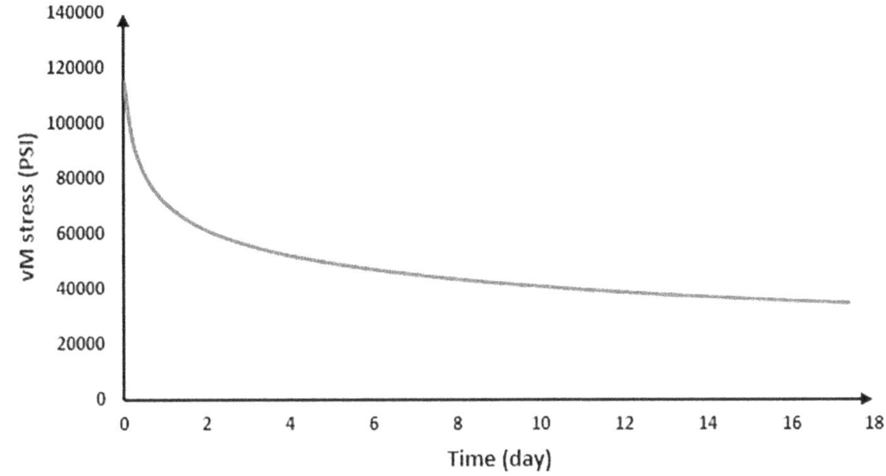

FIGURE 4.16 von-Mises stress of one element of the bolt with time.

4.2.2.7 Summary

The creep of a bolt with pretension was simulated in ANSYS190, which shows the effect of creep.

4.3 DISCUSSION OF VISCOPLASTICITY AND CREEP

The viscoplasticity models such as Perzyna, Peirce, and EVH make the yield stress a function of a plastic strain rate. Thus, a higher plastic strain rate causes a higher yield stress. On the other hand, creep strain is a function of stress, strain, and temperature. It is similar to viscoplasticity, where higher stress levels give higher creep strain rates. A key difference between viscoplasticity and creep is that creep has no yield function; thus, the material is creeping when any stress, strain, or temperature occurs.

REFERENCES

[1] Young, J.F., Mindess, S., Gray, R.J., and Bentur, A., *The Science and Technology of Civil Engineering Materials*, Prentice Hall, Upper Saddle River, NJ, 1998.
[2] ANSYS190 Help Documentation in the help page of product ANSYS190.
[3] Lange, K., *Handbook of Metal Forming*, McGraw-Hill, Inc, New York, 1985.
[4] Lin, R.C., Betten, J., and Brocks, W., Modeling of finite strain viscoplasticity based on the logarithmic corotational description, *Archive of Applied Mechanics*, Vol. 75, 2006, pp. 693–708.
[5] Meyers, M., Chawla, K., *Mechanical Behavior of Materials*, Cambridge University Press, Cambridge, UK, 2008.

5 Influence of Temperature on Material Properties

Chapter 5 focuses on the relationship between temperature and material properties, including the definition of temperature dependency of material properties and simulation of a combustion chamber.

5.1 TEMPERATURE DEPENDENCY OF MATERIAL PROPERTIES

Material properties of metals/alloys are strongly temperature-dependent [1]. By lowering the temperature, the number and tightness of chemical bonds increase. Thus, yielding becomes more and more difficult, and the initial yield stress σ_0 increases. Other plastic parameters have similar changes with temperature. In ANSYS, the temperature dependency is defined using the command TBTEMP. Here is one example of the Chaboche model with different temperatures:

```
TB,CHAB,1,N,
TBTEMP,T1
TBDATA,1,  σ0(T1), C1(T1), γ1(T1), ···
TBTEMP,T2
TBDATA,1,  σ0(T2), C1(T2), γ1(T2), ···
...
TBTEMP,TN
TBDATA,1,  σ0(Tn), C1(Tn), γ1(Tn), ···
```

Temperature affects the mechanical behavior of metals/alloys in many aspects. The temperature directly causes the thermal strain. Also, as mentioned at the beginning of this section, the material properties of the metals/alloys are functions of temperature. Normally, with an increase in temperature, Young's modulus and yield stress decrease. Section 5.2 simulates the combustion chamber to illustrate these effects.

5.2 SIMULATION OF COMBUSTION CHAMBER UNDER DIFFERENT TEMPERATURES

Here, the combustion chamber was simulated in ANSYS190 to study its mechanical behavior with temperature loading.

5.2.1 Finite Element Model

A quarter of the combustion chamber was modeled using Plane182 with kyopt(3)=2 (plane strain condition) (see Figure 5.1).

DOI: 10.1201/9781003436317-6

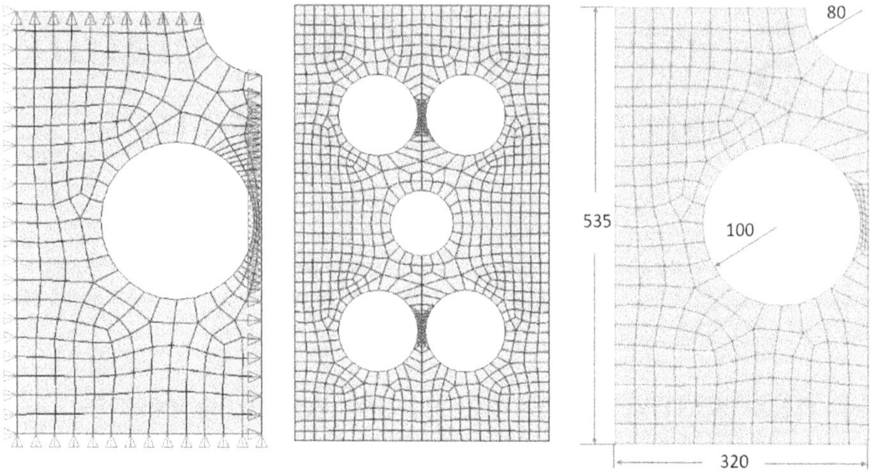

FIGURE 5.1 Finite element model of a combustion chamber. (a) A quarter of the model. (b) Full model. (c) Geometry (all dimensions in mm).

5.2.2 Material Properties

The material of the combustion chamber was defined with the one-term Chaboche model plus the Perzyna option. Also, the material properties of the combustion chamber vary with temperatures (Table 5.1) [2]. These material models were defined in ANSYS as follows:

```
! DEFINE YOUNG'S MODULUS AND POISSON'S RATIO
TB,ELAS,1,3
TBTEMP,20
TBDATA,1,76000,0.33
TBTEMP,100
TBDATA,1,72000,0.33
TBTEMP,300
TBDATA,1,60000,0.33
! DEFINE CHABOCHE MODEL
TB,CHAB,1,4,1
TBTEMP,20
TBDATA,1,80,21700,155
TBTEMP,100
TBDATA,1,63,22000,200
TBTEMP,200
TBDATA,1,52,8460,310
TBTEMP,300
TBDATA,1,4,3780,420
! DEFINE PERZYNA MODEL
TB,RATE,1,3,,PERZYNA                    ! RATE TABLE
TBTEMP,20
```

TABLE 5.1
Material Parameters of the Combustion Chamber [2]

Temperature (°)	20	100	200	300
E (MPa)	76,000	72,000		60,000
	Chaboche Model			
σ_0 (MPa)	80	63	52	4
C_1	21,700	22,000	8,460	3,780
γ_1	155	200	310	420
	Perzyna			
m_1	1/8.5		1/8.5	1/8.5
γ_1	1e-4		3e-6	1e-10
	MISO			
σ_0 (MPa)	80	63	52	4
$\sigma(\varepsilon_{pl} = 0.1)$ (MPa)	200	157	130	10

```
TBDATA,1,1/8.5,1E-4
TBTEMP,200
TBDATA,1,1/8.5,3E-6
TBTEMP,300
TBDATA,1,1/8.5,1E-10
MP,ALPX,1,6E-6

! DEFINE MISO
TB,PLAS,1,4,,MISO
TBTEMP,20
TBPT,DEFI,0,80
TBPT,DEFI,0.1,200
TBTEMP,100
TBPT,DEFI,0,63
TBPT,DEFI,0.1,157
TBTEMP,200
TBPT,DEFI,0,52
TBPT,DEFI,0.1,130
TBTEMP,300
TBPT,DEFI,0,4
TBPT,DEFI,0.1,10
```

5.2.3 Loadings and Boundary Conditions

The combustion chamber was fixed on all sides. In addition, symmetrical conditions were applied on the edges (see Figure 5.1a). The temperature of the combustion chamber changes with time (see Figure 5.2).

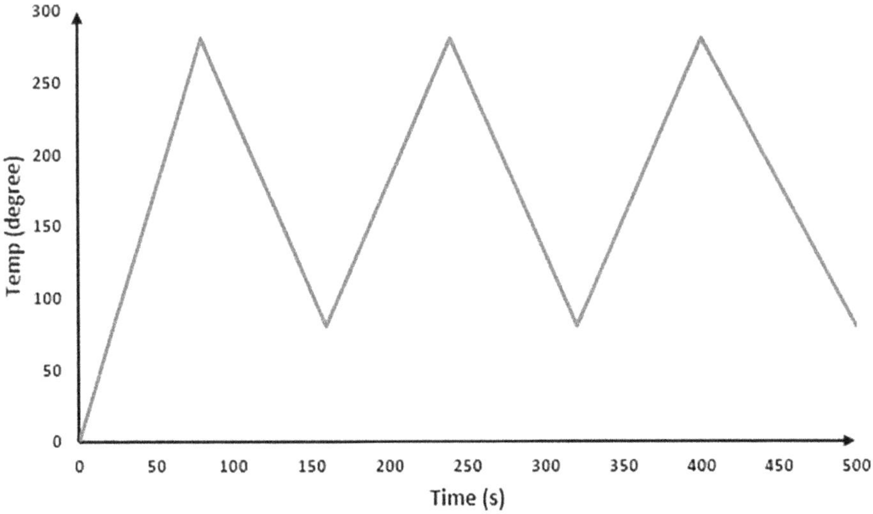

FIGURE 5.2 Loading history of temperature.

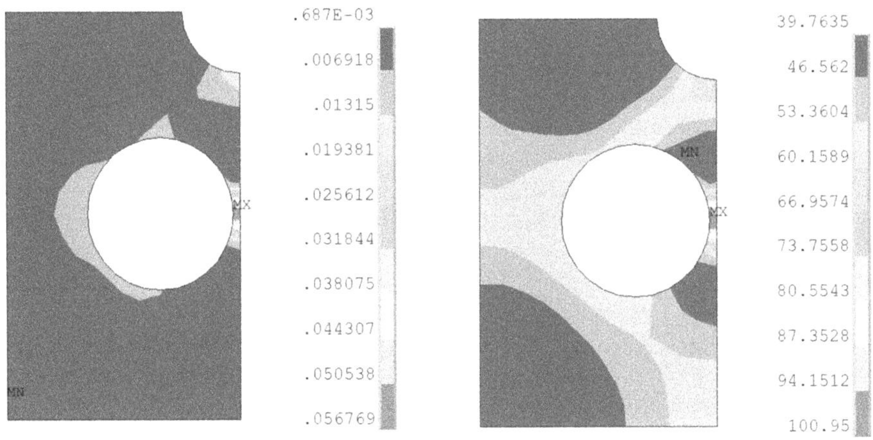

FIGURE 5.3 von-Mises plastic strains and stresses of the combustion chamber at time 80s. (a) von-Mises plastic strains. (b) von-Mises stresses (MPa).

5.2.4 Results

The von-Mises plastic strain and stress of the combustion chamber at each loading step are plotted in Figures 5.3–5.8, respectively. The maximum values occur at the hole edge with the narrowest width and vary at different loading steps. At that location, the time history of the stress, plastic strain, and thermal strain are plotted in Figure 5.9.

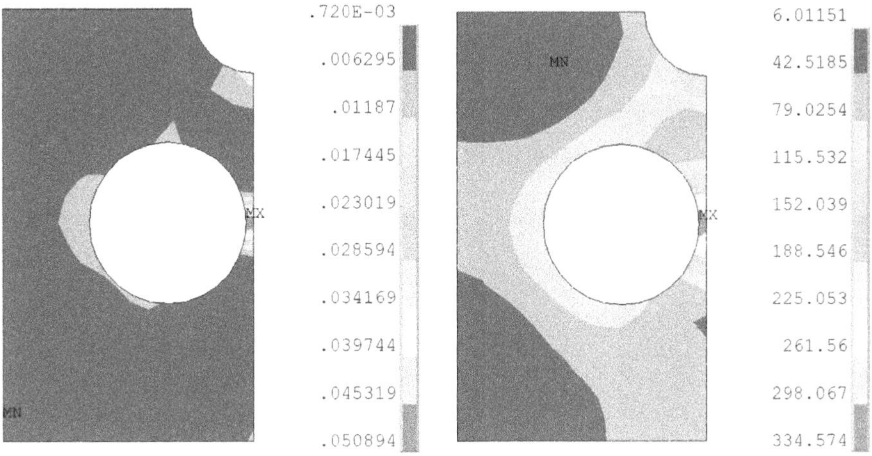

FIGURE 5.4 von-Mises plastic strains and stresses of the combustion chamber at time 160s. (a) von-Mises plastic strains. (b) von-Mises stresses (MPa).

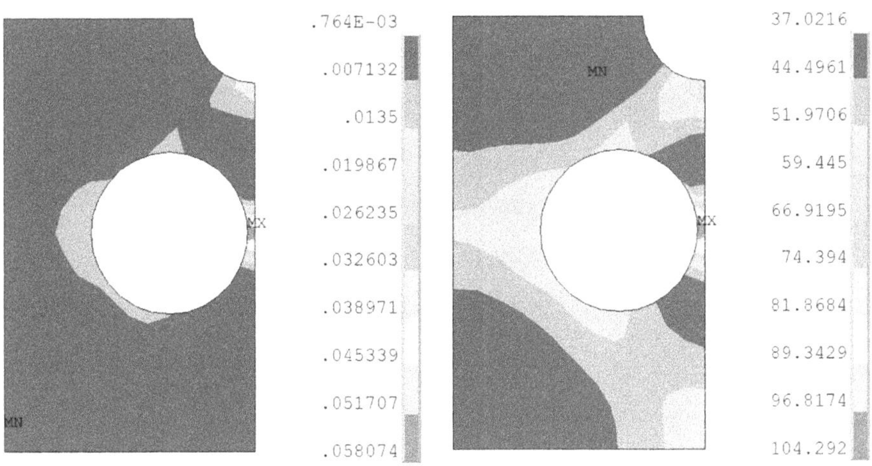

FIGURE 5.5 von-Mises plastic strains and stresses of the combustion chamber at time 240s. (a) von-Mises plastic strains. (b) von-Mises stresses (MPa).

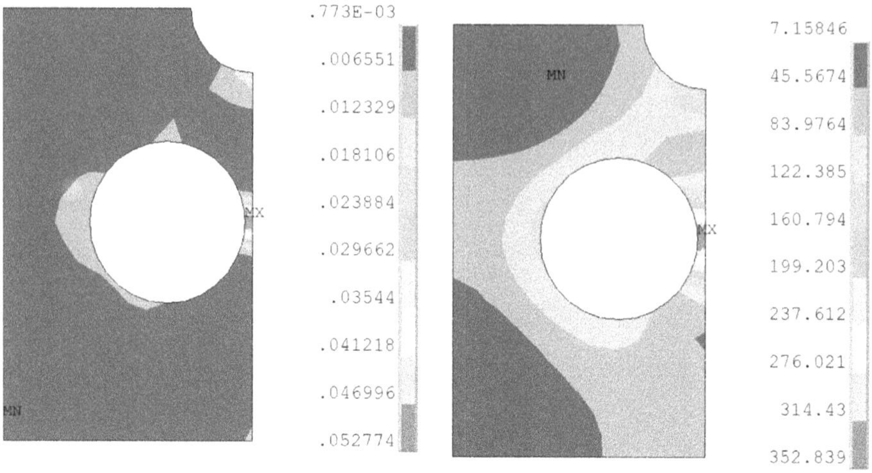

FIGURE 5.6 von-Mises plastic strains and stresses of the combustion chamber at time 320s. (a) von-Mises plastic strains. (b) von-Mises stresses (MPa).

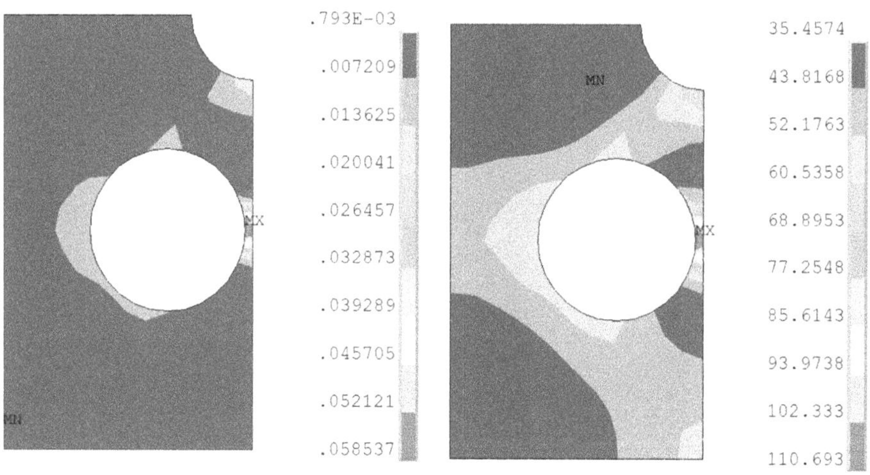

FIGURE 5.7 von-Mises plastic strains and stresses of the combustion chamber at time 400s. (a) von-Mises plastic strains. (b) von-Mises stresses (MPa).

5.2.5 Discussion

A combustion chamber was simulated at different temperatures with material properties as a function of temperature. Obviously, the temperature plays a major role in the model. The temperature causes thermal strain, which directly affects the stresses and strains of the combustion chamber. At the first loading step, the thermal strain caused by the temperature change makes the stress and elastic strain rise quickly. After reaching yield (point a in Figure 5.9c), the plastic strain increases as well. Because the high temperature corresponds to a low yield stress, the stress drops at the

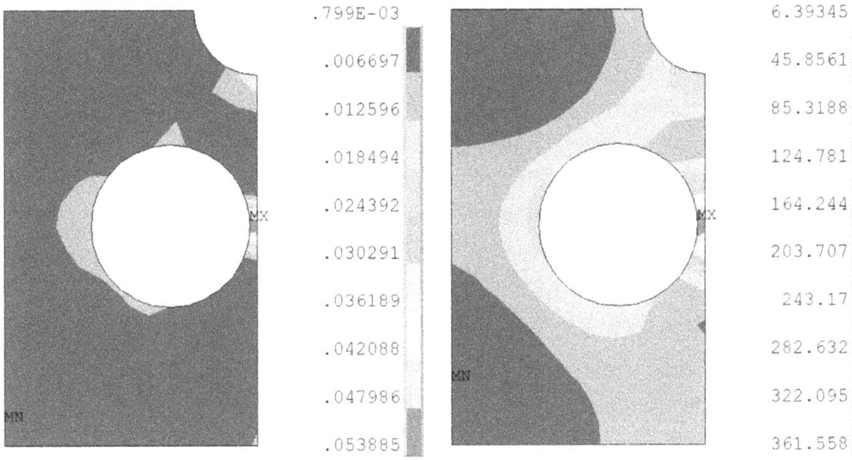

FIGURE 5.8 von-Mises plastic strains and stresses of the combustion chamber at time 500s. (a) von-Mises plastic strains. (b) von-Mises stresses (MPa).

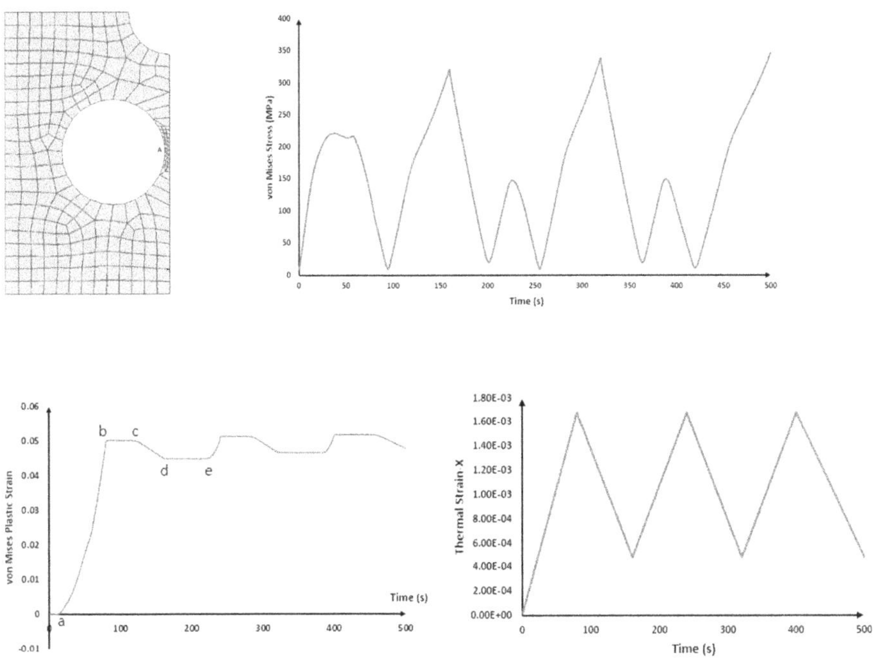

FIGURE 5.9 Results of element A with time. (a) Element A. (b) von-Mises stress of element A with time. (c) von-Mises plastic strain of element A with time. (d) Thermal strain of element A with time.

later stage of the first loading step, although the plastic strain keeps rising. After the temperature decreases in the second loading step, the material is in the elastic stage between points b and c in Figure 5.9c. After reaching the yield surface, the plastic strain changes between points c and d in Figure 5.9c until the end of the second loading step. The same behavior of the stresses and strains in the combustion chamber occurs at the later loading steps. It is also observed that at the beginning, from o to point a in Figure 5.9c, the material is in the elastic stage. With the rise of temperature, Young's modulus decreases, which explains why the stress in this period increases nonlinearly (see Figure 5.9b), although the material is in the elastic stage. Overall, temperature plays two roles in the model: (1) thermal strain and (2) material properties as a function of temperature.

5.2.6 Summary

A combustion chamber was simulated at different temperatures. The computational results indicate that temperature affects the mechanical behavior of the combustion chamber in two aspects: (1) thermal strain and (2) material properties as a function of temperature.

REFERENCES

[1] Roylance, D., *Mechanical Properties of Materials*, MIT, Cambridge, 2008.
[2] https://z-mat.com/PDFs/info/z-mat-detail.pdf

6 Subroutine UserMat

The ANSYS material library provides various material models that cover such fields as mechanical engineering, biomechanics, and geomechanics. However, customers sometimes must develop their own material models in ANSYS if they are not able to find one appropriate for their projects in the library. Currently, ANSYS supports this population by offering UserMat, UserHyper, UserCreep, and UserMatTh [1] as ways to develop their own material subroutines. Since UserMat and UserHyper are the most popular, this chapter focuses on UserMat development, while Chapter 12 emphasizes the UserHyper subroutine.

6.1 USERMAT DEVELOPMENT

UserMat development is problem-dependent. Whether it is straightforward or complicated, it always meets the same requirements: the output of stress $\boldsymbol{\sigma}$ and the material Jacobian matrix $\frac{\partial \sigma}{\partial \varepsilon}$. Here is one UserMat example of linear elasticity.

The stress of the linear elasticity can be computed directly by

$$\sigma = \mathbf{D}\varepsilon \tag{6.1}$$

and its material Jacobian matrix is defined as

$$\frac{\partial \sigma}{\partial \varepsilon} = \mathbf{D} \tag{6.2}$$

Thus, the subroutine involves the following steps:

1. update strain $\boldsymbol{\varepsilon}_{n+1} = \boldsymbol{\varepsilon}_n + \Delta\varepsilon$
2. compute stress $\boldsymbol{\sigma}_{n+1} = \mathbf{D}\boldsymbol{\varepsilon}_{n+1}$
3. output stress $\boldsymbol{\sigma}_{n+1}$ and the material Jacobian matrix $\mathbf{D}$

Following similar steps, the UserMat of the strain-hardening material model was built; it is the focus of the next section.

6.2 USERMAT OF STRAIN-HARDENING MATERIAL MODEL

The strain-hardening material models have been developed to model thermo-mechanical states of metals and alloys and have gained more and more attention when studying the thermo-mechanical states in technological processes of metals and alloys.

6.2.1 Introduction of Strain-Hardening Model

In 1945, Hollomon suggested one widely used strain-hardening equation (see Figure 6.1) [2]:

$$\sigma = K_H \varepsilon_{pl}^{n_H} \tag{6.3}$$

where K_H and n_H are material constants and ε_{pl} is the plastic strain.

Therefore, the stress-strain relationship becomes

$$\varepsilon = \varepsilon_e + \varepsilon_{pl} = \frac{\sigma}{E} + \left(\frac{\sigma}{K_H}\right)^{\frac{1}{n_H}} = \frac{\sigma}{E} + K_s\sigma^{n_s} \tag{6.4}$$

where E is Young's modulus, $K_s = \dfrac{1}{K_H{}^{\frac{1}{n_H}}}$ is the material constant, and $n_s = \dfrac{1}{n_H}$ is the material parameter that describes the hardening behavior of the material.

6.2.2 Implementation UserMat of Strain-Hardening Material Model

The way to implement UserMat of the strain-hardening material model is to simplify it as a one-dimensional problem. In a one-dimensional problem, $\dfrac{d\sigma}{d\varepsilon}$ can be determined from the following equation (6.5):

$$\frac{d\sigma}{d\varepsilon} = \frac{E}{K_s n_s \sigma^{n_s - 1} E + 1} \tag{6.5}$$

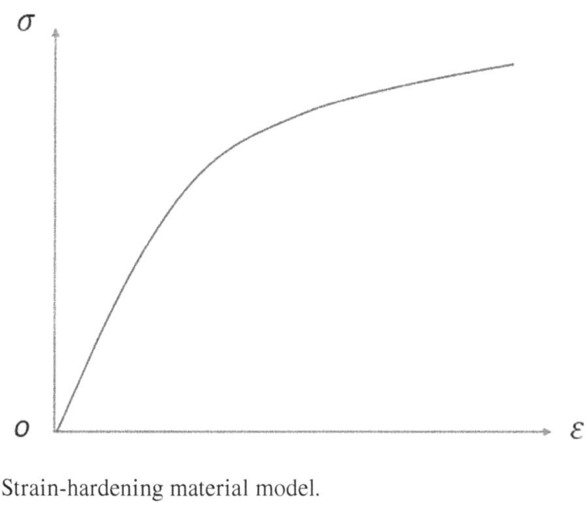

FIGURE 6.1 Strain-hardening material model.

The right side of the above equation can be regarded as an equivalent Young's modulus.

Thus, the implementation of UserMat for the strain-hardening material model involves the following steps:

1. convert three-dimensional strain and stress to one-dimensional equivalent strain and stress;
2. solve one-dimensional nonlinear equation (6.4) using the Newton-Raphson method;
3. calculate the equivalent Young's modulus with the obtained σ in step 2;
4. compute the three-dimensional stress increment with the equivalent Young's modulus ($\Delta\boldsymbol{\sigma}_{eq} = \mathbf{D}_{eq}\Delta\boldsymbol{\varepsilon}$);
5. obtain the true three-dimensional stress increment using scaling from the known one-dimensional stress increment;

$$\Delta\boldsymbol{\sigma} = \Delta\boldsymbol{\sigma}_{eq}\frac{\Delta\sigma_{1D}}{\Delta\sigma_{eq1D}} \tag{6.5}$$

6. compute the three-dimensional stress:

$$\boldsymbol{\sigma}_{n+1} = \boldsymbol{\sigma}_n + \Delta\boldsymbol{\sigma} \tag{6.6}$$

7. output the above three-dimensional stress $\boldsymbol{\sigma}_{n+1}$ and $\mathbf{D}_{eq}$. Also, save the three-dimensional stress $\boldsymbol{\sigma}_{n+1}$ as state variables that can be used in the next iteration.

Following the above steps, the UserMat of the strain-hardening model was completed.

The APDL commands to define the material model in the input file are listed as follows:

```
TB,USER,1,4
TBDATA,1, E, V, KS, NS
! SAVE THE CURRENT STRESSES FOR THE NEXT ITERATION
! THE STATE VARIABLES MUST BE DEFINED. OTHERWISE, IT HAS
! CONVERGENCE ISSUE.
TB,STATE,1,6
```

6.2.3 Validation

A uniaxial test with the developed UserMat was conducted. The results (see Figure 6.2) match the theoretical results well, which validates the developed UserMat.

Furthermore, a $6 \times 1 \times 1$ block with one end fixed and another end loaded with a torque was simulated to test the robustness of the developed UserMat (see Figure 6.3). The solution was completed quickly, and the final von-Mises stress plot looked reasonable (see Figure 6.4).

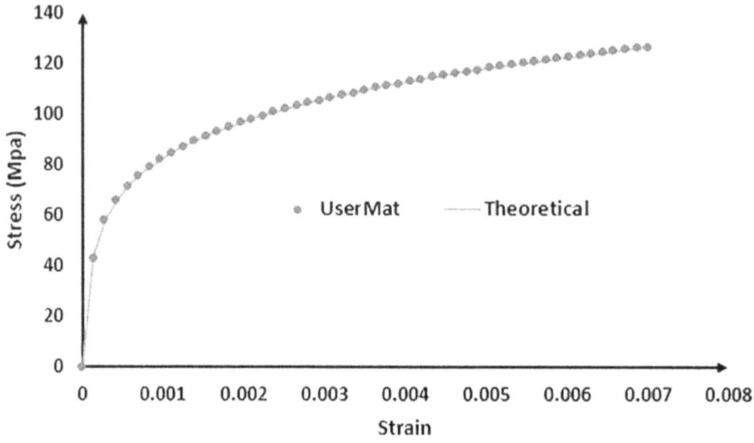

FIGURE 6.2 Uniaxial test results of UserMat.

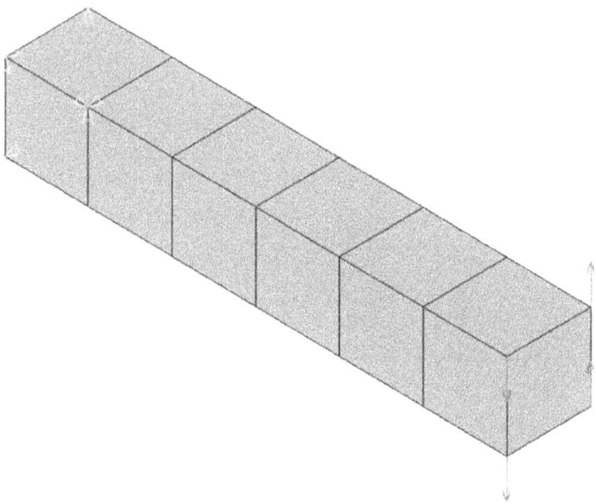

FIGURE 6.3 Finite element model of the performance test.

6.2.4 Summary

The UserMat implementation in ANSYS was introduced with the implementation of the strain-hardening model as an example.

REFERENCES

[1] ANSYS190 Help Documentation in the help page of product ANSYS190.
[2] Hollomon, J.H., Tensile deformation, *Transactions of the Metallurgical Society of AIME*, Vol. 162, 1945, pp. 268–290.

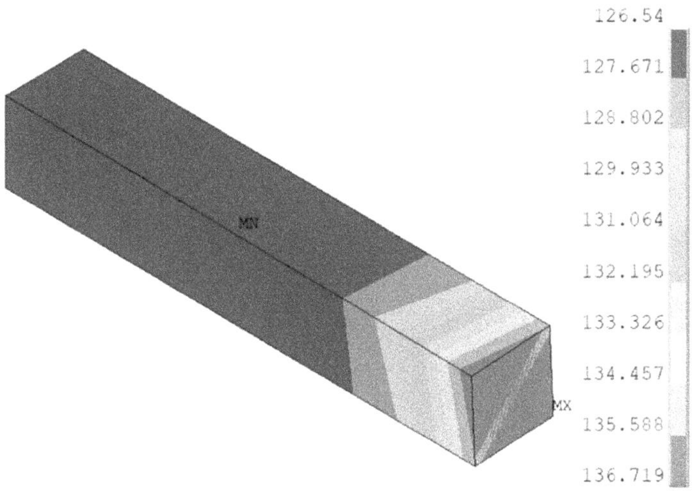

FIGURE 6.4 von-Mises stresses of the performance test (MPa).

Part II

Polymers

Polymers with a molecular structure exhibit large deformation, creep, plastic flow, viscoelastic flow, rupture, and wear. To capture the behaviors of polymers, some hyperelastic material models are available in ANSYS and are discussed in Part II.

Chapter 7 first presents the structure and features of polymers. Chapter 8 then focuses on some hyperelastic models and curve-fitting of their material parameters. Chapter 9 emphasizes the viscoelasticity of polymers and their shift functions. Chapter 10 discusses eight-chained-based viscoplasticity models. The elastomer damage Mullins effect is examined in Chapter 11. Chapter 12 exhibits a simple and useful way to model hyperelastic materials using the subroutine UserHyper in ANSYS.

DOI: 10.1201/9781003436317-8

7 Structure and Features of Polymer

Chapter 7 briefly introduces the structure and features of polymers.

7.1 STRUCTURE OF POLYMER

A polymer is defined as a material consisting of many macromolecules. Each macromolecule is a long chain linked by a number of monomer units formed by covalent bonds. Thus, a polymer is formed with three different length scales. In the first scale, the most local scale, the atoms are arrayed into monomer units and joined together by covalent bonds. The other two scales are the long chain-like structure and the network structure. The different chain molecules have three primary forms (see Figure 7.1) [1]: (1) linear macromolecule, (2) branched macromolecule, and (3) crosslinked macromolecule.

Different structures lead to various material properties. The linear macromolecule becomes soft when heated and hard when cooled, which is called thermoplasticity. On the other hand, the crosslinked macromolecule cannot flow or melt easily with heat because its relative motion among the molecular chains is limited; that is referred to as thermoset.

7.2 FEATURES OF POLYMER

Compared with metal, polymers have the following features:

1. The network structures interact with weak van der Waals' forces, resulting in low stiffness and high ductility of polymers. The elastic modulus of polymers is about 1e6 times less than that of metal.
2. The Poisson's ratio of polymer is around 0.49, close to 0.50, which means polymers have a very high bulk modulus.

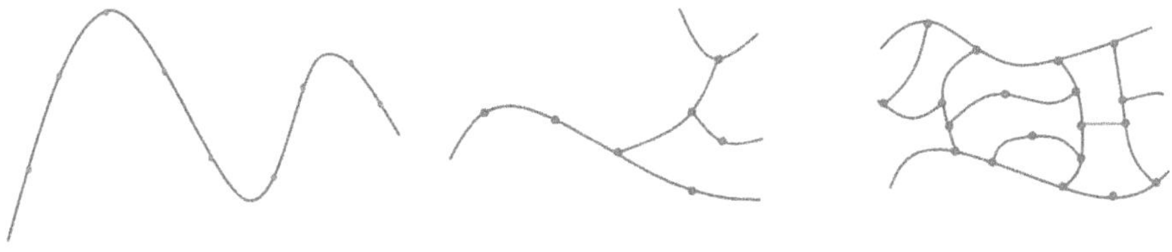

FIGURE 7.1 Structures of a single macromolecular chain. (a) Linear macromolecule. (b) Branched macromolecule. (c) Crosslinked macromolecule.

DOI: 10.1201/9781003436317-9

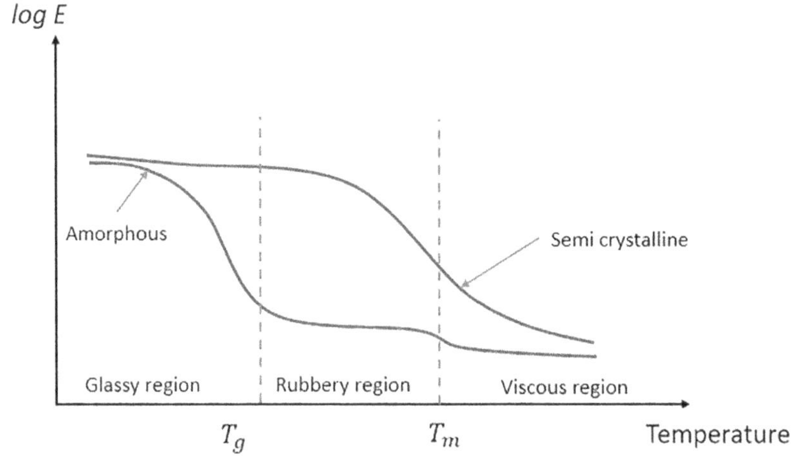

FIGURE 7.2 Variety of Young's modulus with temperature for different classes of elastomers [2]. T_g, glass transition temperature; T_m, melting temperature.

3. Uncrosslinked polymers reveal time-dependent elastic deformation, namely relaxation.
4. Temperature influences all aspects of the mechanical behavior of polymers. Figure 7.2 plots variations of the material's stiffness (Young's modulus) with temperature [2].

REFERENCES

[1] Zhang, S., *Tribology of Elastomers*, Elsevier, Amsterdam, 2004.
[2] Bergstrom, J., *Mechanics of Solid Polymers*, Elsevier, Amsterdam, 2015.

8 Hyperelasticity

After presenting some common hyperplastic models, Chapter 8 then conducts the curve-fitting of material parameters.

8.1 SOME WIDELY USED HYPERELASTIC MODELS

8.1.1 Neo-Hookean Model

The Neo-Hookean model is a simple hyperelastic model with two material parameters: a shear modulus μ and a bulk modulus k. The elastic energy density of the Neo-Hookean model is expressed as [1]

$$W = \frac{\mu}{2}(I_1 - 3) + \frac{1}{d}(J - 1)^2 \tag{8.1}$$

where d – material incompressibility parameter, $= \frac{2}{k}$.

The equation is a linear function of I_1. Thus, it cannot catch the large-strain nonlinear behavior of many elastomers. In addition, no I_2 term in the energy equation makes stress prediction low when biaxial loading dominates. Overall, the advantage of the Neo-Hookean model is that the model is robust and computationally efficient; its limitation is that it cannot give an accurate prediction in case of a large strain.

8.1.2 Mooney-Rivlin Model

Unlike the Neo-Hookean model, the Mooney-Rivlin model includes a linear term of I_2. The two-parameter Mooney-Rivlin model has the following form [1]:

$$W = c_{10}(I_1 - 3) + c_{01}(I_2 - 3) + \frac{1}{d}(J - 1)^2 \tag{8.2}$$

where c_{10} and c_{01} are material parameters.

The Mooney-Rivlin model may improve the stress prediction, but it also raises another issue – having a negative c_{01} term, which may cause the model to be unstable in the finite deformation.

8.1.3 Yeoh Model

The Yeoh model surpasses the Neo-Hookean model because it has a high order of I_1 without I_2 [1]:

DOI: 10.1201/9781003436317-10

$$W = \sum_{i=1}^{N} c_{i0} (I_1 - 3)^i + \sum_{k}^{N} \frac{1}{d_k} (J - 1)^{2k} \tag{8.3}$$

Therefore, the Yeoh model can predict stress more accurately than the Neo-Hookean model, and it can also avoid the stability issue of the Mooney-Rivlin model.

8.1.4 Polynomial Model

The polynomial model – a generalization of the Neo-Hookean model, the Mooney-Rivlin model, and the Yeoh Model – involves taking a polynomial expansion in terms of I_1 and I_2 as [1]

$$W = \sum_{i+j=1}^{N} c_{ij} (I_1 - 3)^i (I_2 - 3)^j + \sum_{k}^{N} \frac{1}{d_k} (J - 1)^{2k} \tag{8.4}$$

The disadvantage of the polynomial model is that it is very difficult to determine the material parameters to make an accurate and robust computation in multiaxial loading cases.

8.1.5 Gent Model

The Gent model, an extension of the Neo-Hookean model, aims to capture the large-strain behavior of elastomer-like material expressed as [1]

$$W = -\frac{\mu J_m}{2} \ln\left(1 - \frac{I_1 - 3}{J_m}\right) + \frac{1}{d}\left(\frac{J^2 - 1}{2} - lnJ\right) \tag{8.5}$$

where J_m is the limiting value of $I_1 - 3$.

In the above equation, the Gent model has one extra parameter J_m to control the limited chain extensibility at large strains.

8.1.6 Ogden Model

The Ogden model differs from the other hyperelastic materials because the energy density of Ogden is expressed in terms of the applied principal stretches [1]:

$$W = \sum_{i=1}^{N} \frac{\mu_i}{\alpha_i} \left(\bar{\lambda}_1^{\alpha_i} + \bar{\lambda}_2^{\alpha_i} + \bar{\lambda}_3^{\alpha_i} - 3\right) + \sum_{k}^{N} \frac{1}{d_k} (J - 1)^{2k} \tag{8.6}$$

where λ_i $(i = 1,2,3)$ are the principal stretches.

Like the polynomial model, the Ogden model also makes it difficult to determine the material parameters to give an accurate and stable prediction.

8.1.7 Arruda-Boyce Model

The Arruda-Boyce model is based on the eight-chain model that is derived from the deformation behavior of the microstructure of elastomers. The elastic energy density of the Arruda-Boyce model is expressed in terms of μ, limiting network stretch λ_L, and d [1]

$$W = \mu \begin{bmatrix} \frac{1}{2}(I_1 - 3) + \frac{1}{20\lambda_L^2}\left(I_1^2 - 9\right) + \frac{11}{1{,}050\lambda_L^4}\left(I_1^3 - 27\right) \\ + \frac{19}{7{,}000\lambda_L^6}\left(I_1^4 - 81\right) + \frac{519}{673{,}750\lambda_L^8}\left(I_1^5 - 243\right) \end{bmatrix} + \frac{1}{d}\left(\frac{J^2 - 1}{2} - lnJ\right) \quad (8.7)$$

where λ_L is the limiting network stretch.

The Arruda-Boyce model is more accurate than the Neo-Hookean model and Mooney-Rivlin model and is nearly as accurate as the Yeoh model. However, it underpredicts the biaxial response because it has no I_2.

Table 8.1 summarizes the predictive capabilities of the listed seven common hyperelastic models, in which the accuracy of the different models is defined by the coefficient of determination (R^2) [2]. The three-term Ogden performs the best; next is the Yeoh. The Arruda-Boyce model, Gent model, and two-term Ogden model are very close behind the first two models. The Neo-Hookean model is the least accurate one because it has the simplest form. The Mooney-Rivlin model is slightly better than the Neo-Hookean model.

8.2 STABILITY DISCUSSION

Some hyperelastic models are not always stable. For example, the Gent model under the uniaxial tension in the large strain gets the stress reduced to even negative (see Figure 8.1). Therefore, Drucker's stability should be checked to examine if the model is stable [2]:

$$\Delta(J\boldsymbol{\sigma}) : \Delta(\mathbf{E}_{ln}) \geq 0 \quad (8.8)$$

where $J = \det[\mathbf{F}]$, $\boldsymbol{\sigma}$ is the Cauchy stress, and $\mathbf{E}_{ln}$ is the logarithmic strain.

TABLE 8.1
Comparison between the Predictive Capabilities of Various Hyperelasticity Models [2]

	Neo-Hookean	Mooney-Rivlin	Yeoh	Arruda-Boyce	Ogden (2-term)	Ogden (3-term)	Gent
R^2 –Prediction	0.794	0.843	0.980	0.973	0.977	0.998	0.972

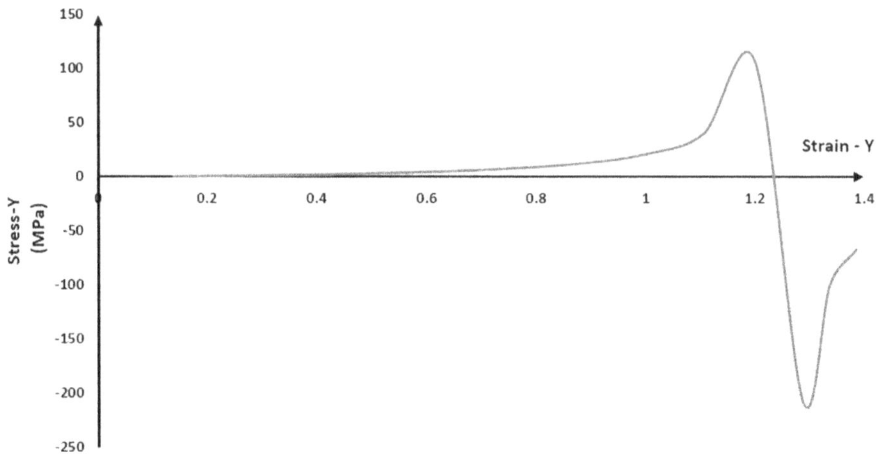

FIGURE 8.1 Stress-strain curve of the Gent model under tension.

Moreover, Drucker's stability should be checked for all loading conditions because some hyperelastic materials may be Drucker's stable in tension but unstable in shear loading. Therefore, in practice, the examination of Drucker's stability of a hyperelastic model is to check the stability of the model under a set of common loading conditions like uniaxial, biaxial, and simple shear. If a material model passes these tests, it may be regarded as relatively safe to use.

8.3 CURVE-FITTING OF MATERIAL PARAMETERS FROM EXPERIMENTAL DATA

A curve-fitting tool of hyperelasticity is available in ANSYS190. The commands are the same as those for the Chaboche model in Section 3.2.2, including input of experimental data, definition of the material model, initialization of coefficients, and solution. Here is one example of the curve-fitting of the two-term Ogden model:

```
/PREP7
! DEFINE UNIAXIAL DATA
TBFT, EADD, 1, UNIA, CFHY-OG-U.EXP
TBFT, EADD, 1, BIAX, CFHY-OG-B.EXP
! DEFINE MATERIAL: TWO-TERM OGDEN MODEL
TBFT, FADD, 1, HYPER, OGDEN,2
! SET THE INITIAL VALUE OF COEFFICIENTS
TBFT,SET,1,HYPER,OGDE,2,1,1
TBFT,SET,1,HYPER,OGDE,2,2,1
TBFT,SET,1,HYPER,OGDE,2,3,1
TBFT,SET,1,HYPER,OGDE,2,4,1

! DEFINE SOLUTION PARAMETERS IF ANY
TBFT, SOLVE, 1, HYPER, OGDEN,2, 1, 500
/OUT
! PRINT THE RESULTS
TBFT, LIST, 1
```

```
TBFT,FSET,1,HYPER,OGDEN,2
TBLIS,ALL,ALL
```

The experimental data of rubber for curve-fitting include uniaxial and biaxial tensions [2,3]. For comparison, the curve-fitting of material parameters was conducted for the Arruda-Boyce model, the three-parameter Mooney-Rivlin model, and the two-term Ogden model. The curve-fitting results are plotted in Figure 8.2, and the obtained material parameters are listed as follows:

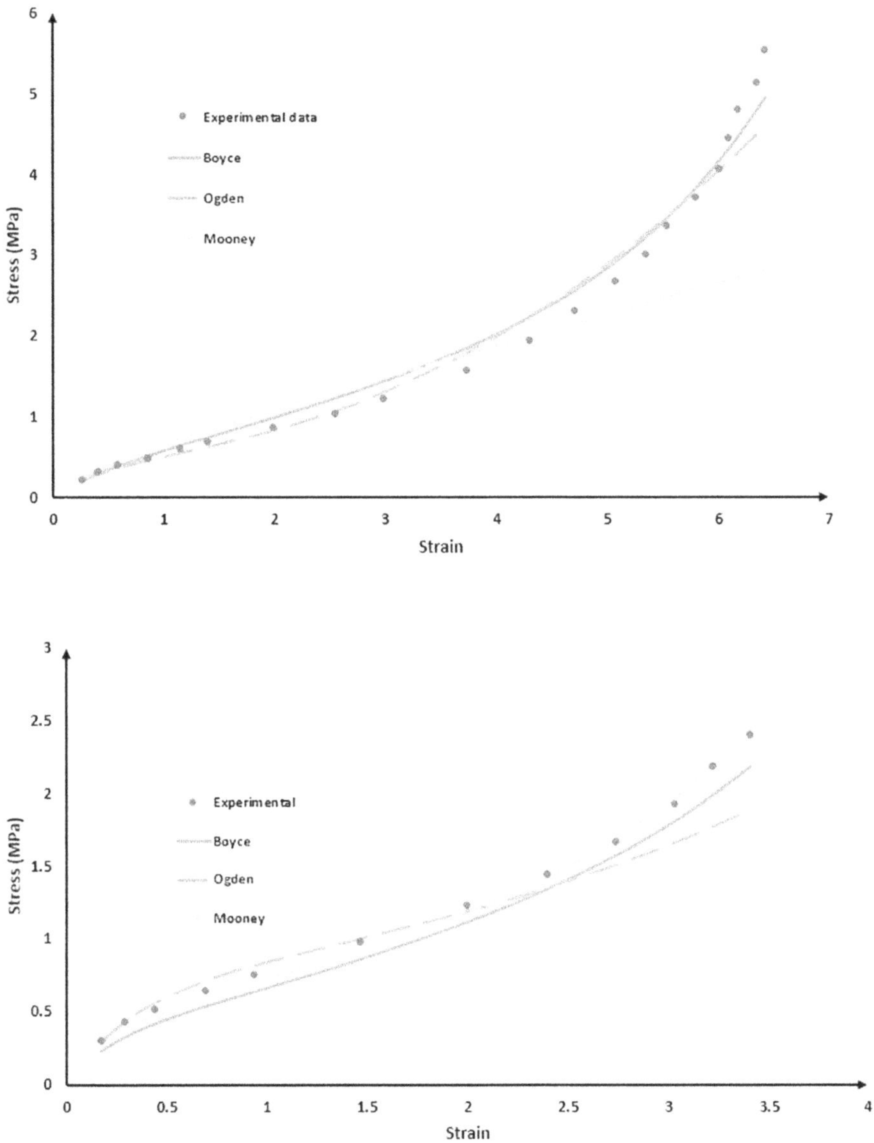

FIGURE 8.2 Curve-fitting results of three hyperelastic models. (a) Uniaxial tension. (b) Biaxial tension.

Arruda-Boyce model: $\mu = 0.3148$, $\lambda_L = 5.032$
Mooney-Rivlin model: C_{10} = 1.89e-1, C_{01} = −1.933e-3, C_{11} = 1.183e-4
Ogden model: μ_1 = 3.565e-2, α_1 = 3.398, μ_2 = 1.272e3, α_2 = 4.948e-4

Figure 8.2 indicates that both the Arruda-Boyce model and Ogden model fit the experimental data of the uniaxial tension, while the fitting results of the Mooney-Rivlin model in the uniaxial tension at the large strains are far from the experimental data, which are consistent with the predictive capabilities of various polymers in Table 8.1. For biaxial tension, the Mooney-Rivlin model fits well because the strain energy function of the Mooney-Rivlin model has I_2 term, while both the Arruda-Boyce and Ogden models are a little different from the experimental data.

REFERENCES

[1] ANSYS190 Help documentation in the help page of product ANSYS190.
[2] Bergstrom, J., *Mechanics of Solid Polymers*, Elsevier, Amsterdam, 2015.
[3] Treloar, L.R., *The Physics of Rubber Elasticity*. Oxford University Press, Oxford, 1975.

9 Viscoelasticity of Polymers

Chapter 9 first examines the viscoelasticity of polymers, including various viscoelastic models and their application for study of the viscoelasticity of liver soft tissues. It then discusses different shift functions and simulates the stress evolution of a cooling glass tube.

9.1 VISCOELASTICITY OF POLYMERS

Polymers exhibit viscoelasticity. For example, when a car runs on the highway, its tires get heated, which is due to the energy dissipation by the viscoelasticity of the tires converted to heat. From the molecular aspect, viscoelasticity is regarded as a molecular rearrangement [1]. When a polymer is loaded with forces, some of the long polymer chains change positions. This rearrangement is called creep. When these long polymer chains move to accompany the loaded forces, the polymers remain a solid material. Thus, a back stress occurs in the material. When the back stress reaches the same level as the applied forces, the creep of the material stops. If the applied loadings are removed, the remaining back stresses cause the polymer to return to its original shape. Some viscoelastic polymers include amorphous polymers, semi-crystalline polymers, and biopolymers.

9.1.1 Linear Viscoelastic Models

A viscoelastic material has an elastic part and a viscous part. The elastic part can be modeled as springs, and its stress-strain relationship is expressed as

$$\sigma = E\varepsilon \tag{9.1}$$

The viscous part can be regarded as dashpots, and its stress is determined by its time derivative of strain

$$\sigma = \eta \frac{d\varepsilon}{dt} \tag{9.2}$$

where η is the viscosity of the material.

Some models, such as the Maxwell model, the Kelvin-Voigt model, and the Burgers model, have been developed to predict the response of viscoelastic material under different loading conditions [2]. These models contain diverse combinations of springs and dashpots, which are briefly introduced as follows.

9.1.1.1 Maxwell Model

In the Maxwell model, a viscous damper and an elastic spring are connected in series (see Figure 9.1). The model is described by the following equation:

DOI: 10.1201/9781003436317-11

FIGURE 9.1 Maxwell model.

$$\sigma + \frac{\eta}{E}\dot{\sigma} = \eta\dot{\varepsilon} \tag{9.3}$$

The above equation indicates that under a constant strain, the stresses decay gradually. Thus, the Maxwell model accurately predicts the stress relaxation of the polymers. However, the equation (9.3) shows that under a constant stress, the strain increases linearly with time as long as the force is loaded, which goes against the mechanical behavior of most polymers. Therefore, the Maxwell model cannot be used for modeling creep.

9.1.1.2 Kelvin-Voigt Model

The Kelvin-Voigt model can be represented by a viscous damper and an elastic spring connected in parallel (see Figure 9.2). The model is expressed by

$$\sigma = E\varepsilon + \eta\dot{\varepsilon} \tag{9.4}$$

The Kelvin-Voigt model shows that under a constant stress, the deformation rate of the material reduces gradually until it reaches a steady state. After the stress is removed, the material gradually returns to its undeformed state. Thus, unlike the Maxwell model, the Kelvin-Voigt model can effectively predict creep. Yet, it models stress relaxation much less accurately than the Maxwell model.

9.1.1.3 Burgers Model

To overcome the limitations of the Maxwell and Kelvin-Voigt models, the Burgers model was created to combine these two models in a series (see Figure 9.3). The constituent relationship is given as

$$\sigma + \left(\frac{\eta_1}{E_1} + \frac{\eta_2}{E_1} + \frac{\eta_2}{E_2}\right)\dot{\sigma} + \frac{\eta_1}{E_1}\frac{\eta_2}{E_2}\ddot{\sigma} = \eta_2\dot{\varepsilon} + \frac{\eta_1\eta_2}{E_1}\ddot{\varepsilon} \tag{9.5}$$

The above second-order differential equation of the Burgers model can be used to describe the viscoelastic behaviors of the polymers under different loadings.

9.1.1.4 Generalized Maxwell Model

The relaxation of the polymers has a varying time distribution because shorter molecular segments contribute less than longer ones. Therefore, the Generalized Maxwell model has many spring-dashpot Maxwell elements that consider this factor (see Figure 9.4).

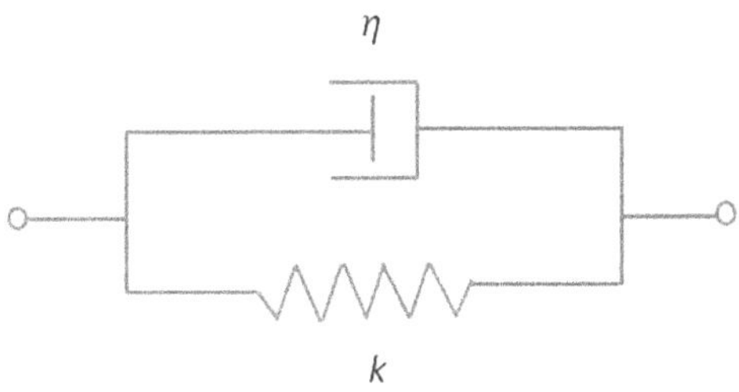

FIGURE 9.2 Kelvin-Voigt model.

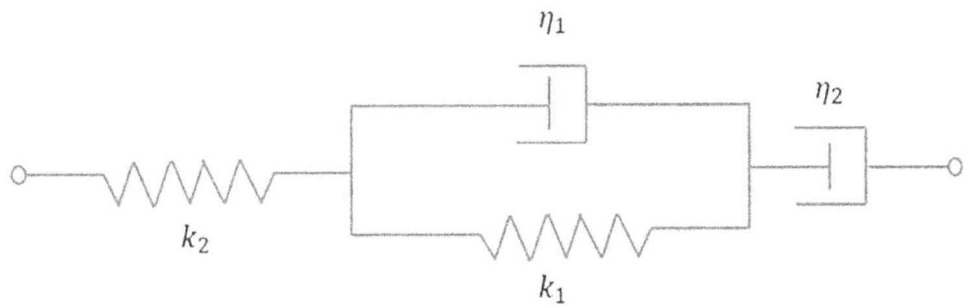

FIGURE 9.3 Burgers model.

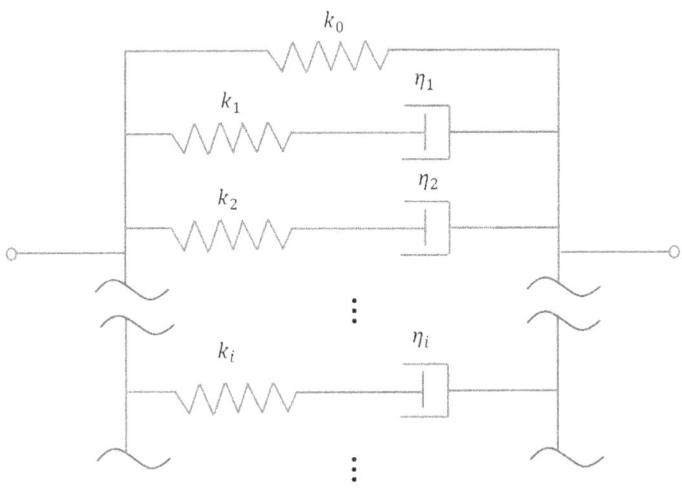

FIGURE 9.4 Generalized Maxwell model.

A Generalized Maxwell model in three dimensions is described by the Prony series

$$\boldsymbol{\sigma} = \int_0^t 2G(t-\tau)\frac{d\mathbf{e}}{d\tau}d\tau + \mathbf{I}\int_0^t K(t-\tau)\frac{d\Delta}{d\tau}d\tau \tag{9.6}$$

where $G(t)$ – the Prony series shear-relaxation moduli, expressed as

$$G(t) = G_0\left[\alpha_\infty^G + \sum_{i=1}^{n_G}\alpha_i^G \exp\left(-\frac{t}{\tau_i^G}\right)\right] \tag{9.7}$$

and $K(t)$ – the Prony series bulk-relaxation moduli, specified by

$$K(t) = K_0\left[\alpha_\infty^K + \sum_{i=1}^{n_K}\alpha_i^K \exp\left(-\frac{t}{\tau_i^K}\right)\right] \tag{9.8}$$

G_0 , K_0 – relaxation moduli at t=0;
n_G , n_K – number of Prony terms;
α_i^G, α_i^K – relative moduli;
τ_i^G , τ_i^K – relaxation time;
σ – Cauchy stress;
$\mathbf{e}$ – deviatoric strain;
Δ – volumetric strain;
τ – time; and
$\mathbf{I}$ – identity tensor.

In ANSYS, Prony series constants are specified by [3]

```
MP, EX, 1, E                          ! define Young's modulus
MP, NUXY, 1, ν                        ! define Poisson's ratio
TB, PRONY,,, NG, SHEAR                ! the shear Prony data table
TBDATA,1, α1G , τ1G , ··· αnG , τnG ,
TB, PRONY,,, NK, BULK                 ! the bulk Prony data table
TBDATA,1, α1K , τ1K , ··· αnK , τnK ,
```

Next, the Prony series was applied to study the viscoelasticity of liver soft tissues.

9.1.2 Simulation of Viscoelasticity of Liver Soft Tissues

In vehicle crashes, the liver is the most often injured abdominal organ [4,5]. According to the past study, the liver was always modeled with hyperelastic materials [6–11]. This study, however, simulated the viscoelasticity of the liver under compression in ANSYS190.

9.1.2.1 Finite Element Model

The liver was placed on a plate and compressed by a cylinder on top. The whole model was meshed with Plane182 in plane strain condition (see Figure 9.5).

9.1.2.2 Material Properties

The liver was modeled with the one-term Mooney-Rivlin model, and its viscoelasticity was simulated with the Prony series. The material parameters [12], which are listed in Table 9.1, were defined by ANSYS APDL commands as

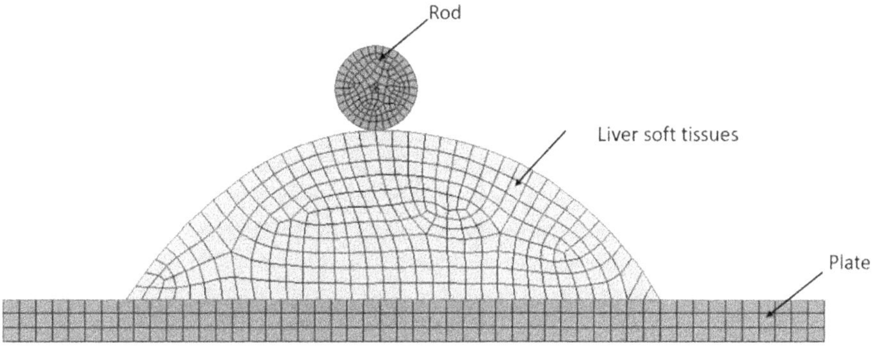

FIGURE 9.5 Finite element model of liver soft tissues.

TABLE 9.1
Material Parameters of the Liver [13]

	α_n^G	τ_n^G(s)
1	0.288	0.58
2	0.171	11.07
3	0.226	140.48

```
TB, HYPE, 1, 1, 1, Mooney  ! Mooney-Rivlin MODEL
TBDATA, 1, 21.76e-3*0.542,-21.76e-3*0.542/2
TB, PRONY, 1, 1, 3, SHEAR  ! PRONY'S SERIES
TBDATA, 1, 33.92/117.96, 0.58, 20.16/117.96, 11.07,
26.61/117.96, 140.48
```

9.1.2.3 Contact Definition

The contact between the plate and the liver and the contact between the liver and the cylinder were assumed to be standard contacts. The cylinder was regarded as rigid and controlled by a pilot node.

9.1.2.4 Loadings and Boundary Conditions

The loading is divided into two steps. In the first step, the cylinder moved down 0.5 mm in 0.1 second. In step two, it held 1000s. The bottom of the plate was fixed in the vertical direction (see Figure 9.6).

9.1.2.5 Results

The von-Mises stresses of the model at the end of the first and second steps are presented in Figures 9.7 and 9.8, respectively. The stresses are localized around the loading area, and the maximum stress decrease was around 65% between the first and second steps. Figure 9.9 illustrates the time history of the reaction force, which decreases with time and then approaches a constant after 400 s.

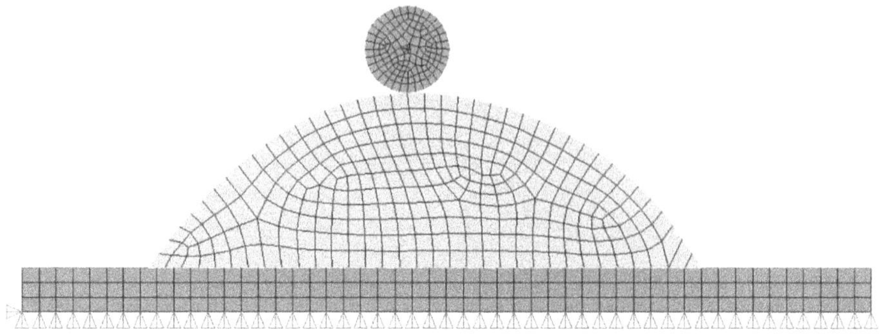

FIGURE 9.6 Boundary conditions.

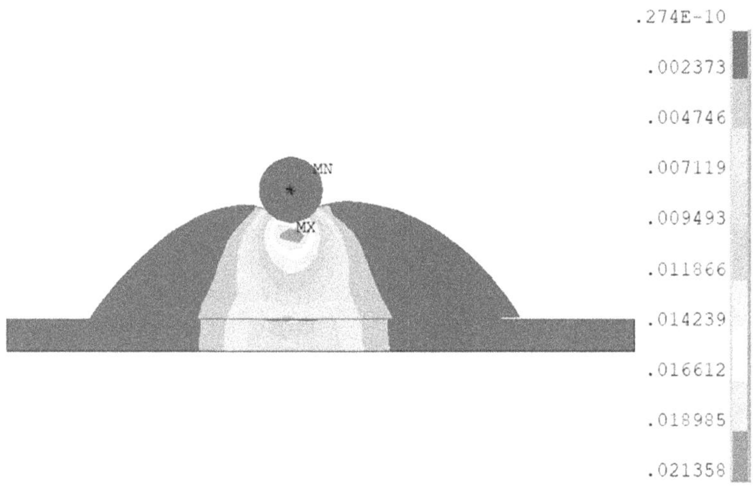

FIGURE 9.7 von-Mises stresses of soft tissues at the end of the first step ($t = 0.1$ s) (MPa).

9.1.2.6 Discussion

A 2D finite element model of the liver was built to study its viscoelastic response under the compression of a cylinder. The reaction force decreases with time and approaches a constant after a certain time, which is because the elastic Mooney-Rivlin model controls its stress response in the first loading step, and the Prony series governs its viscoelastic behavior in the second loading step. Without a definition of the Prony series in the material model, the stress state and reaction force should not change in the second step.

The drawback of this study is that the liver was simplified as a plane strain state, which significantly differs from the true liver. Future studies should use a 3D model of the liver with the geometry from a CT scan.

9.1.2.7 Summary

The viscoelasticity of the liver under compression of a cylinder was studied in ANSYS190. The stress and reaction force decrease with time, which reflects the viscoelasticity of the liver.

FIGURE 9.8 von-Mises stresses of soft tissues at the end of the second step ($t = 1000\,s$) (MPa).

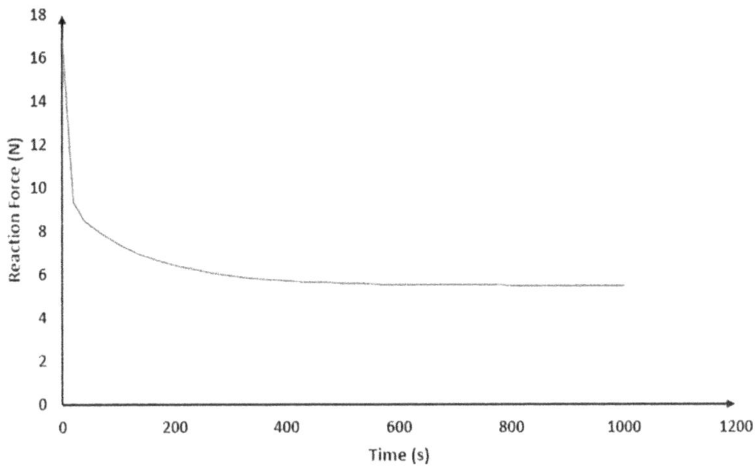

FIGURE 9.9 Reaction force with time.

9.2 SHIFT FUNCTIONS

The stress relaxation time τ in the viscoelasticity of the polymers is not constant but is strongly dependent on the temperature. The material is assumed to be thermorheologically simple (TRS), which means that the change in temperature can be transformed into a time scale change by shifting the relaxation time using a shift function [14]. Some shift functions have been developed to model TRS.

9.2.1 Some Shift Functions

Some shift functions, such as Williams-Landel-Ferry and Tool-Narayanaswamy, are introduced as follows [3]:

1. Williams-Landel-Ferry (WLF) shift function

$$\log_{10}(a_T) = \frac{C_1(T - T_r)}{C_2 + (T - T_r)} \tag{9.9}$$

where T is the temperature, Tr is a reference temperature, and C_1 are C_2 are material constants.

WLF shift function is defined in ANSYS by

```
TB, SHIFT,,,,WLF
TBDATA,1, Tr, C1 , C2
```

2. Tool-Narayanaswamy shift function

$$\ln(a_T) = \frac{H}{R}\left(\frac{1}{T_r} - \frac{1}{T}\right) \tag{9.10}$$

where H is the activation energy and R is the universal gas constant.

In ANSYS, the Tool-Narayanaswamy shift function is specified by

```
TB, SHIFT,,,,TN
TBDATA, 1, Tr, H/R
```

3. Tool-Narayanaswamy shift function with fictive temperature

$$\ln(a_T) = \frac{H}{R}\left(\frac{1}{T_r} - \frac{X}{T} - \frac{1-X}{T_f}\right) \tag{9.11}$$

where X – weight, and

T_f – fictive temperature and determined by

$$T_f = \sum_{i=1}^{n} C_{fi} T_{fi} \tag{9.12}$$

where n is the number of partial fictive temperatures, C_{fi} is the fictive temperature relaxation coefficient, and T_{fi} is the partial fictive temperature.

In ANSYS, Tool-Narayanaswamy shift function with fictive temperature is defined by

```
TB, SHIFT,,,,FICT
TBDATA, 1, Tr, H/R, X
TBDATA, 4, Tf1, Cf1, τf1
......
TBDATA, 3n+1, Tfn, Cfn, τfn
! glass thermal expansion coefficient
TBDATA, 3n+4, αg0, αg1, αg2, αg3, αg4
! liquid thermal expansion coefficient
TBDATA, 3n+9, αl0, αl1, αl2, αl3, αl4
TBDATA, 3n+14, Tr
```

After the shift function a_T is defined, the stress relaxation time τ is scaled as

$$\tau = \frac{\tau_0}{a_T} \tag{9.13}$$

where τ_0 is the stress relaxation time at $T = T_r$

Next, a glass tube from above glass transition temperature to below glass transition temperature was simulated using the shift function.

9.2.2 Simulation of Stress Evolution of a Glass Tube

The stress evolution of a 30 mm thick glass tube was studied when its temperature decreased from above glass transition temperature to below glass transition temperature [15].

9.2.2.1 Finite Element Model

A 2D axisymmetrical glass tube was modeled in ANSYS190 using CPT212 with keyopt (3)=1 (axisymmetrical) and keyopt(11)=1 (thermal DOF) (see Figure 9.10). Also, the model was simplified using the symmetrical condition in the vertical direction.

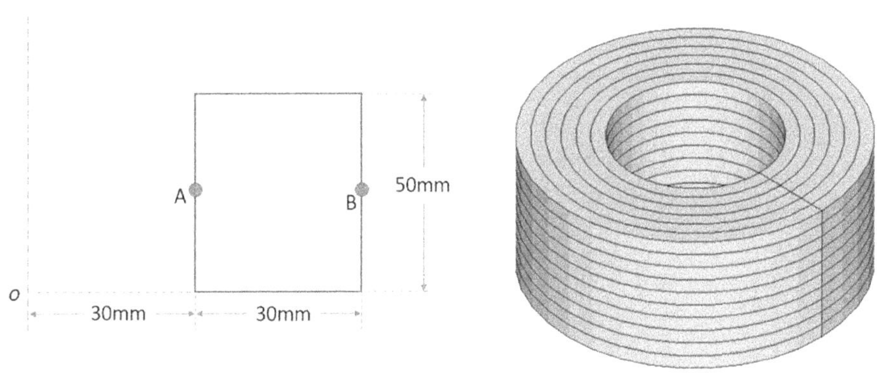

FIGURE 9.10 Geometry of the glass tube: (a) 2D axisymmetric and (b) 3D.

9.2.2.2 Material Properties

Young's modulus and Poisson's ratio of the glass were specified as 72.5 GPa and 0.3, respectively. The Prony series and six terms for structural relaxation are listed in Tables 9.2 and 9.3, respectively.

The material model was defined by the following commands:

```
TB,PRONY,1,3,SHEAR
TBDATA,1,0.422,0.0689,0.423,0.0065,0.155,0.0001

TB,SHIFT,1,6,FICT
TBDATA,1,618,6.45E4,0.53
TBDATA,4,618,0.108,3.0,618,0.443,0.671
TBDATA,10,618,0.166,0.247,618,0.161,0.091
TBDATA,16,618,0.046,0.033,618,0.076,0.008

MPTEMP,1,270
MPTEMP,2,550
MP,CTEX,1,6.47E-6
MP,CTEX,1,34.3E-6
```

9.2.2.3 Loadings and Boundary Conditions

The axisymmetrical condition and the symmetrical condition were applied to edges (see Figure 9.11). The temperature loaded on the inner and outer surfaces changed with time (see Figure 9.12).

TABLE 9.2
Prony Series of Glass Tube [15]

	α_n^G	τ_n^G (s)
1	0.422	0.0689
2	0.423	0.0065
3	0.155	0.0001

TABLE 9.3
Structural Relaxation of Glass Tube [15]

	T_{fi}	C_{fi}	τ_{fi} (s)
1	618	0.108	3.0
2	618	0.443	0.671
3	618	0.166	0.247
4	618	0.161	0.091
5	618	0.046	0.033
6	618	0.076	0.008

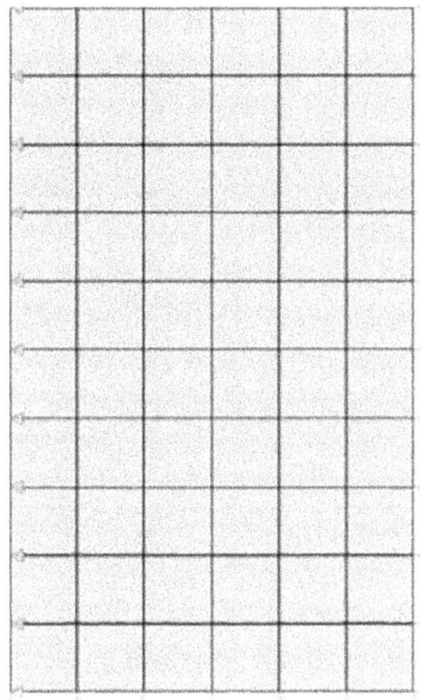

FIGURE 9.11 Boundary conditions.

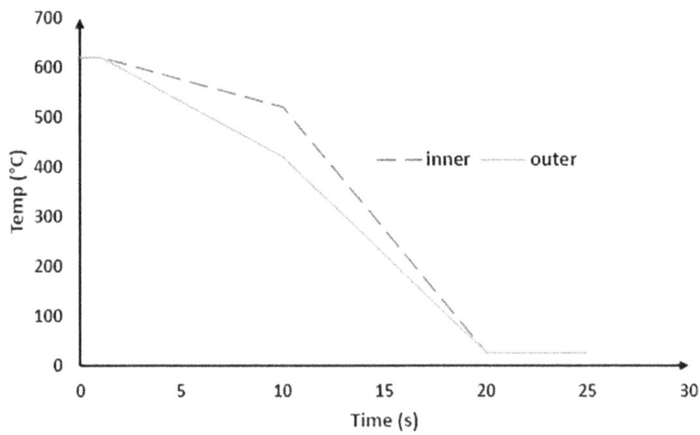

FIGURE 9.12 Temperature change at the inner and outer surfaces.

9.2.2.4 Results

The fictive temperatures at the inner surface A and the outer surface B are plotted in Figure 9.13. As the outer surface is cooled faster than the inner surface, the fictive temperature at the outer surface is higher than that at the inner surface.

It is expected that the axial thermal strains of A and B (see Figure 9.14) are similar to their temperature changes (see Figure 9.12).

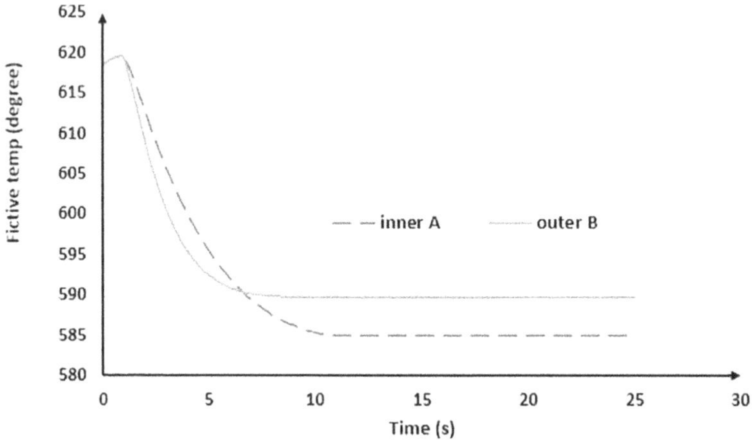

FIGURE 9.13 Fictive temperatures of the inner and outer surfaces.

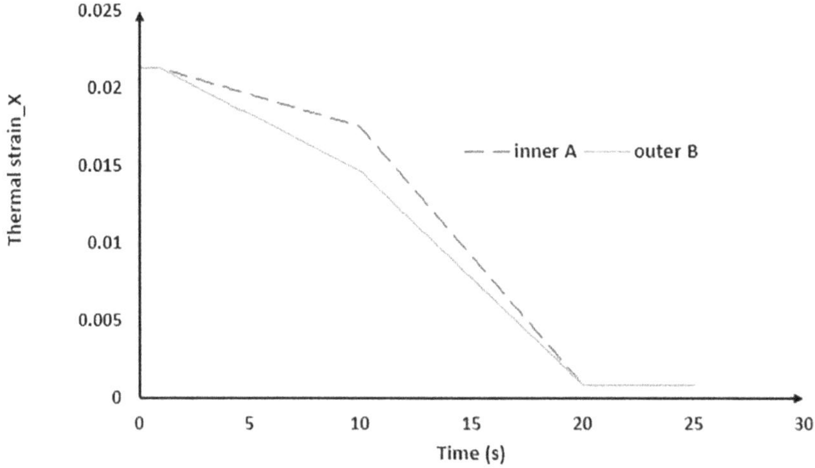

FIGURE 9.14 Axial thermal strain at the inner and outer surfaces.

The axial stresses at the inner surface and outer surface after 10 s follow symmetrical paths that show the inner surface under compression and the outer surface in tension (see Figure 9.15). On the other hand, the hoop stresses at the inner surface and outer surface have opposite results; the hoop stresses at the inner surface are positive and those at the outer surface are negative (see Figure 9.16).

All the above results match the reference [15], which validated the built model.

9.2.2.5 Summary

The stress evolution of a glass tube was solved using the coupled analysis of structure and temperature; the results of fictive temperature, thermal strain, and stress with time match the reference.

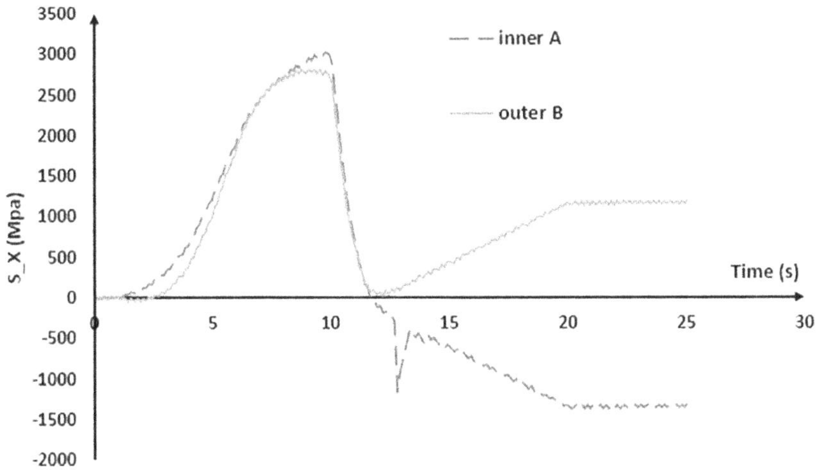

FIGURE 9.15 Axial stresses at the inner and outer surfaces.

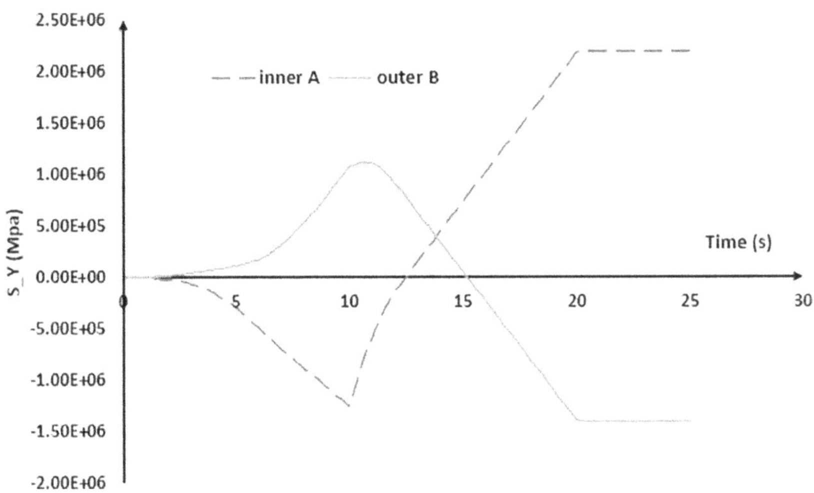

FIGURE 9.16 Hoop stresses at the inner and outer surfaces.

REFERENCES

[1] McCrum, N.G., Buckley, C.P., and Bucknell, C.B., *Principles of Polymer Engineering*, Oxford University Press, Oxford, 2003.
[2] Roylance, D., *Engineering Viscoelasticity*, MIT, Cambridge, 2001.
[3] ANSYS190 Help Documentation in the help page of product ANSYS190.
[4] Elhagediab, A.M., and Rouhana, S., Patterns of abdominal injury in frontal automotive crashes, *16th International ESV Conference Proceedings*, NHTSA, Washington DC, 1998, pp. 770–780.
[5] Rouhana, S.W., and Foster, M.E., Lateral impact-an analysis of the statistics in the NCSS, *Proc. 29th Stapp Car Crash Conference*, 1985, pp. 79–98.

[6] Gao, Z., Lister, K., and Desai, J.P., Constitutive modeling of liver tissue: Experiment and theory, *Annals of Biomedical Engineering*, Vol. 38, 2010, p. 505.
[7] Roan, E., and Vemaganti, K., The nonlinear material properties of liver tissue determined from no-slip uniaxial compression experiments, *Journal of Biomechanical Engineering*, Vol. 129, 2007, pp. 450–456.
[8] Chui, C., Kobayashi, E., and Chen, X., et al., Combined compression and elongation experiments and non-linear modeling of liver tissue for surgical simulation, *Medical & Biological Engineering & Computing*, Vol. 42, 2004, pp. 787–798.
[9] Sakuma, I., Nishimura, Y., and Chui, C., et al., In vitro measurement of mechanical properties of liver tissue under compression and elongation using a new test piece holding method with surgical glue, *Surgery Simulation and Soft Tissue Modeling*, Springer, Berlin Heidelberg, 2003, pp. 284–292.
[10] Fu, Y.B., Chui, C.K., and Teo, C.L., Liver tissue characterization from uniaxial stress-strain data using probabilistic and inverse finite element methods, *Journal of the Mechanical Behavior of Biomedical Materials*, Vol. 20, 2013, pp. 105–112.
[11] Umale, S., Deck, C., and Bourdet, N., et al., Experimental mechanical characterization of abdominal organs: liver, kidney & spleen, *Journal of the Mechanical Behavior of Biomedical Materials*, Vol. 17, 2013, pp. 22–33.
[12] Sato, F., Yamamoto, Y., Ito, D., et al., Hyper-viscoelastic response of perfused liver under dynamic compression and estimation of tissue strain thresholds with a liver finite element model, *IRCOBI Conference*, 2013, pp. 736–750.
[13] Estermann, S.J., Pahr, D.H., and Reisinger, A., Hyperelastic and viscoelastic characterization of hepatic tissue under uniaxial tension in time and frequency domain, *Journal of the Mechanical Behavior of Biomedical Materials*, Vol. 112, 2020, p. 104038.
[14] Scherer, G.W., *Relaxation in Glass and Composites*, John Wiley and Sons, Hoboken, 1986.
[15] Zhang, Z., and Zhang, R., Implementation of a viscoelastic material model to simulate relaxation in glass transition, *The COMSOL Conference*, Boston, 2014.

10 Eight-Chain-Based Viscoplasticity Models

The eight-chain model, also called as the Arruda-Boyce model, is widely used to simulate rate-independent rubber-like materials [1]. The model assumes that the chain molecules are positioned along the diagonals of a unit cell in the principal stretch space (see Figure 10.1). The parameters of this model are physically based on micromechanics of chain molecules.

Using the eight-chain theory, the Arruda-Boyce hyperelastic model was derived to describe the mechanical responses [1]. The Bergstrom-Boyce model followed; it was developed to include the visoplastic part [2]. Then, the three-network model was created for the thermoplastic material [3] (Figure 10.2). This chapter highlights the Bergstrom-Boyce model, one typical viscoplasticity model. Those readers interested in the three-network model should read Bergstrom's book [4].

10.1 BERGSTROM-BOYCE MODEL

The Bergstrom-Boyce model can be represented using two parallel networks, A and B (see Figure 10.2) [2]. A is a nonlinear hyperelastic component, and B is a nonlinear hyperelastic component in a series with a nonlinear viscoelastic flow part. The total stress is expressed by

$$\sigma = \sigma_A + \sigma_B \tag{10.1}$$

$$\sigma_A = \frac{\mu_A}{J\overline{\lambda}^*} \frac{L^{-1}\left(\overline{\lambda}^* \big/ \lambda_A^{lock}\right)}{L^{-1}\left(1 \big/ \lambda_A^{lock}\right)} dev\left[\mathbf{B}^*\right] + \kappa(J-1)\mathbf{I} \tag{10.2}$$

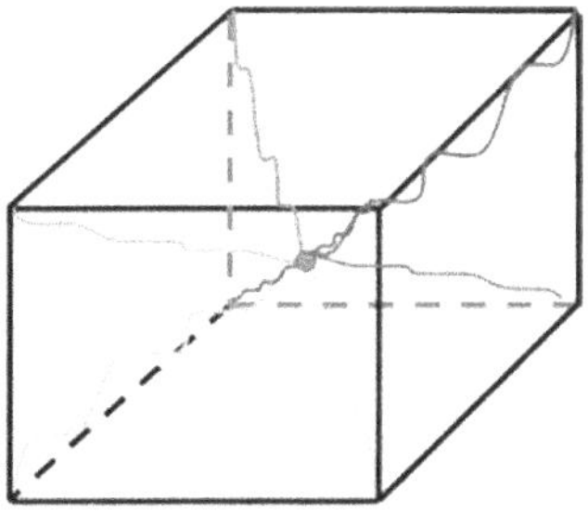

FIGURE 10.1 Eight-chain model.

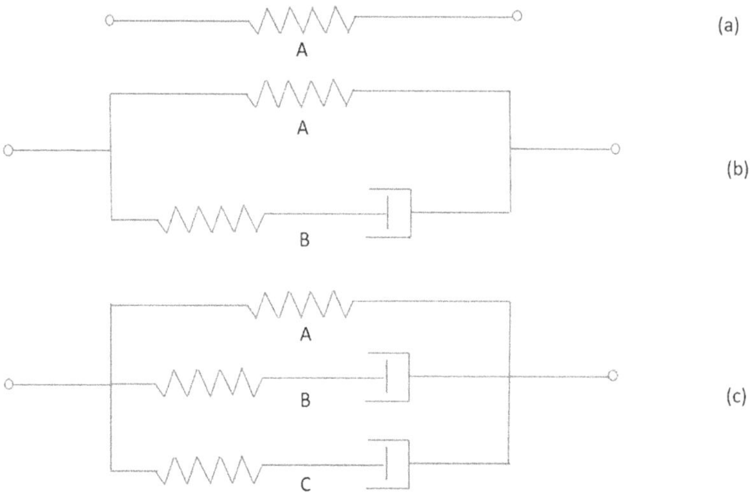

FIGURE 10.2 Models based on the eight-chain model. (a) Arruda -Boyce model, (b) Bergstrom-Boyce model, and (c) three-network model.

$$\sigma_B = \frac{\mu_B}{J_B^e \overline{\lambda_B^{e*}}} \frac{L^{-1}\left(\overline{\lambda_B^{e*}} / \lambda_B^{lock}\right)}{L^{-1}\left(1 / \lambda_B^{lock}\right)} dev\left[\mathbf{B}_B^{e*}\right] + \kappa\left(J_B^e - 1\right)\mathbf{I} \tag{10.3}$$

where μ_A, $\lambda_A{}^{lock}$, μ_B, $\lambda_B{}^{lock}$, κ are material parameters, L^{-1} is the inverse Langevin function, $\mathbf{B}^*$ is the distortional part of left Cauchy-Green tensor, and $\overline{\lambda^{e*}}$ is the chain stretch of network.

The rate-equation for viscous flow is specified by [2]

$$\dot{\gamma}_B^p = \dot{\gamma}_0(\overline{\lambda_B^p} - 1 + \epsilon)^C[\frac{\tau}{\tau_{base}}]^m \tag{10.4}$$

where ϵ, C, τ_{base}, and m are material parameters, $\dot{\gamma}_0$ is a constant to ensure dimensional consistency, and λ_B^p is the viscoelastic chain stretch.

The Bergstrom-Boyce model is defined in ANSYS by the following commands:

TB, BB,,,,ISO
TBDATA,1, μ_A, λ_A^2, μ_B, λ_B^2, $\frac{\dot{\gamma}_0}{\tau_{base}^m}$
TBDATA,6, ϵ, m

TB,BB,,,,PVOL
TBDATA,1,1/κ

The Bergstrom-Boyce model is widely used to predict the nonlinear time-dependent large strain behavior of elastomer-like materials. The application of the Bergstrom-Boyce model led to the simulation of the small punch test (SPT) in the next section.

10.2 SIMULATION OF SMALL PUNCH TEST

The SPT was developed to determine the mechanical behavior of various materials [5]. As it requires minimal material, the SPT emerges as an attractive and commonly used process.

10.2.1 Finite Element Model

The two-dimensional axisymmetrical finite element model of SPT was created following the dimensions of SPT [5] (see Figure 10.3). The specimen was held by the holder, and the die was positioned on the surface of the specimen. The die moved down to make the specimen deformed.

10.2.2 Material Properties

The material was modeled by the Bergstrom-Boyce model with the following material parameters [3]:

```
LAMBDAL = 3.25
KAPPA = 6000    !MPA
MUA = 22.0      !MPA
MUB = 200       !MPA
GM = 5.906e-8
M = 20
TB,BB,1,,,ISO
TBDATA,1,MUA,LAMBDAL*LAMBDAL,MUB,LAMBDAL*LAMBDAL
TBDATA,5,GM,0,M
TB,BB,1,,,PVOL
TBDATA,1,1/KAPPA
```

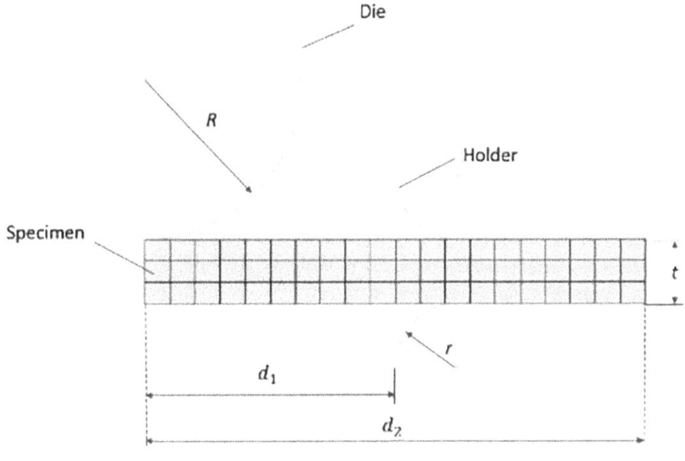

FIGURE 10.3 Finite element model of the small punch test (d_2=4 mm, t=0.5 mm, R=1.25 mm, d_1= 2 mm, r=0.5 mm).

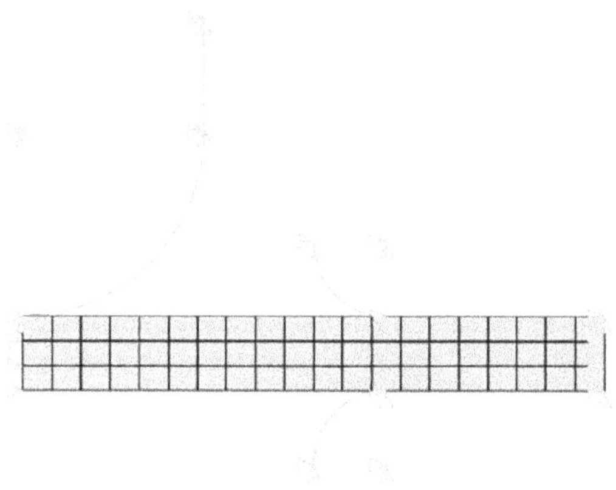

FIGURE 10.4 Boundary conditions.

10.2.3 Contact Definition

The contact between the sheet and the die and the contact between the sheet and the holder were assumed to be standard contacts. The die and the holder were regarded as rigid (see Figure 10.4).

10.2.4 Boundary Conditions

The die moved down 4 mm and the holder was fixed in all degrees of freedom. The right end of the sheet was also fixed (see Figure 10.4).

10.2.5 Solution Setting

The whole time was set at 480 s to meet the requirement of the die loading speed of 0.5 mm/min. In addition, a large deflection effect must be turned on because the specimen has a large deformation.

10.2.6 Results

The final von-Mises stresses of the specimen are plotted in Figure 10.5. The maximum stress occurs at the center, where there is the biggest deformation. Also, the strain as big as 2.06 in Figure 10.6 shows the robustness of the material model. Besides, the reaction force increases with the vertical displacement of the die (see Figure 10.7).

10.2.7 Summary

The SPT was simulated in ANSYS190. The solution converged even with large strain, which shows the robustness of the Bergstrom-Boyce model.

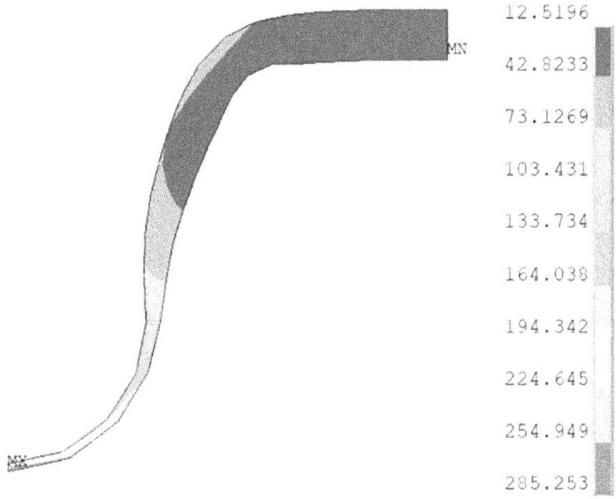

FIGURE 10.5 Final von-Mises stresses of the model (MPa).

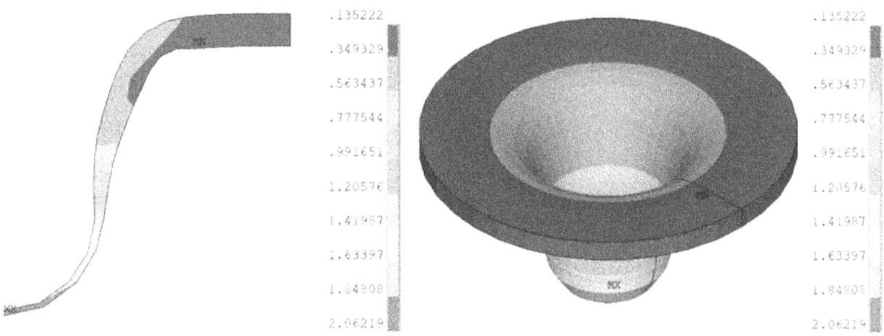

FIGURE 10.6 Final von-Mises strains of the model: (a) 2D axisymmetrical model and (b) 3D model.

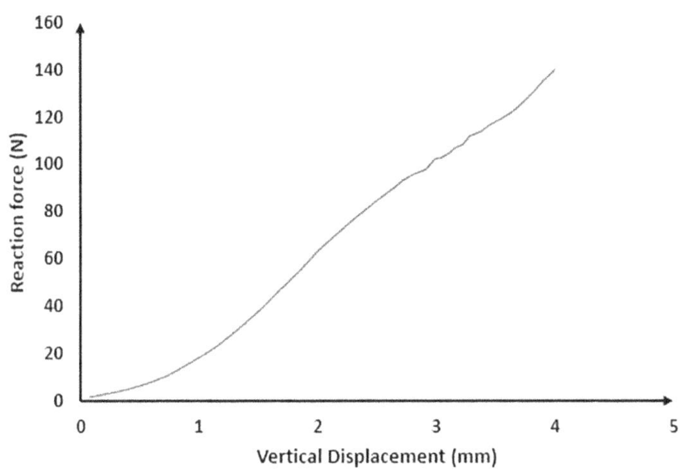

FIGURE 10.7 Reaction forces with vertical displacements.

REFERENCES

[1] Arruda, E., and Boyce, M.C., A three-dimensional constitutional model for the large stretch behavior of rubber elastic materials, *Journal of The Mechanics and Physics of Solids*, Vol. 42, 1993, pp. 389–412.

[2] Bergstrom, J.S., and Boyce, M.C., Constitutive modeling of the time-dependent and cyclic loading of elastomers and application to soft biological tissues, *Mechanics of Materials*, Vol. 33, 2001, pp. 523–530.

[3] Bergstrom, J.S., and Bischoff, J.E., An Advanced Thermomechanical Constitutive Model for UHMWPE, *International Journal of Structural Changes in Solids*, Vol. 2, 2010, pp. 31–39.

[4] Bergstrom, J., *Mechanics of Solid Polymers*, Elsevier, Amsterdam, 2015.

[5] Kuna, M., Abendroth, M., Linse, T., and Rasche, S., Characterization of temperature and irradiation dependent material properties by means of the small punch test and advanced numerical methods, *Metallurgical Journal*, Vol. 63, 2010, pp. 151–160.

11 Mullins Effect

Chapter 11 first introduces the Mullins effect and then focuses on the Ogden-Roxburgh Mullins effect model. The final part of Chapter 11 shows the damage evolution of a rubber tire on the road in ANSYS.

11.1 INTRODUCTION OF MULLINS EFFECT

Significant softening of the elastomers occurs at the first few loading cycles (see Figure 11.1). After that, the material properties remain constant [1–4]. This phenomenon is called the Mullins effect. The Mullins effect applies only to elastomers and elastomer-like materials. From the aspect of the molecular structure, the damage is caused by either a break of molecular chains at the interface of filler particles or a breakdown of filler particle clusters. Experiments find the following features of the Mullins effect:

1. Softening increases with filler particle concentration;
2. In the first few cycles, stress on a given strain is higher than in the following cycles;
3. Damage to the material becomes the highest in the first few cycles and reduces significantly in the following cycles;
4. The maximum applied strain determines the amount of damage. When the damage is caused by a cyclic loading, the material exhibits more damage accumulation if the higher loading is applied in the following cycles; and
5. The Mullins effect can be recovered slowly with time, and temperature significantly affects its recovery rate.

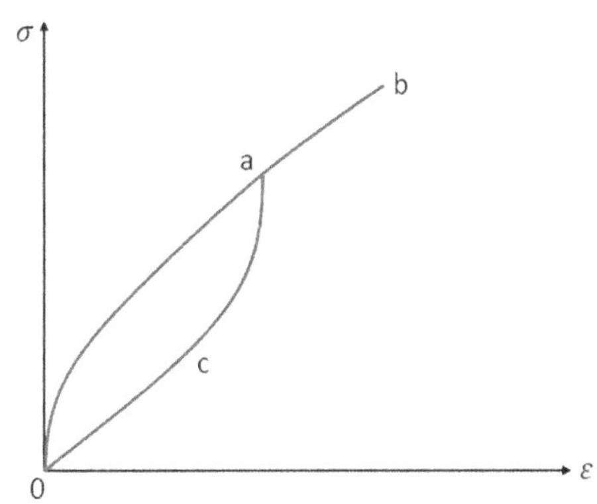

FIGURE 11.1 Response of Mullins effect model. (a) First loading 0→a→b. (b) First unloading b→a→c→0. (c) Second loading 0→c→a.

DOI: 10.1201/9781003436317-13

A few models have been developed to predict the Mullins effect, such as the Ogden-Roxburgh Mullins effect model and Qi-Boyce Mullins effect model [5]. The Ogden-Roxburgh Mullins effect model, which was implemented in ANSYS, is introduced in the following section.

11.2 OGDEN-ROXBURGH MULLINS EFFECT MODEL

In the Ogden-Roxburgh Mullins effect model, an internal variable η is introduced in the Helmholtz free energy density Ψ to track the damage evolution in the material, which is expressed as [6]

$$\eta = 1 - \frac{1}{r}\operatorname{erf}\left(\frac{\psi^{max} - \psi}{m + \beta\psi^{max}}\right) \tag{11.1}$$

where r, β, and m are material constants,

erf (x) – the error function,
ψ – current strain energy density for the material point, and
ψ^{max} – maximum strain energy density for the material point in its loading history.

The Ogden-Roxburgh Mullins effect model can be applied to predict any isotropic hyperelastic material model.

The Mullins effect is defined in ANSYS as

```
TB, CDM,,,,PSE2
TBDATA, 1, r, m, β
```

11.3 SIMULATION OF A RUBBER TIRE WITH THE MULLINS EFFECT

In this study, a tire compressed on a rigid road was modeled in ANSYS with the Mullins effect. The damage caused during the rotation of the tire should be helpful for understanding the physical meaning of the Mullins effect.

11.3.1 Finite Element Model

The model was created to represent a tire on a rigid road that was simplified as a rigid plate (see Figure 11.2). The tire was meshed by SOLID185.

11.3.2 Material Properties

The rubber was simulated with a three-term YEOH material model. In addition, the damage to the rubber was modeled using the modified Ogden-Roxburgh Mullins effect. The corresponding APDL commands are given as

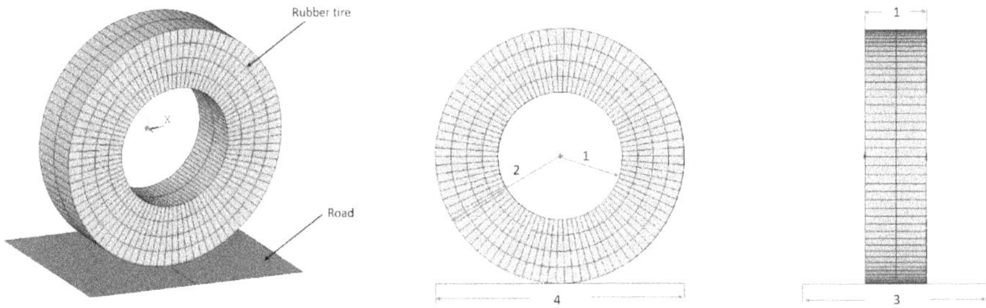

FIGURE 11.2 Finite element model of a rubber tire on the road. (a) Finite element model. (b) Geometry of a mini tire (all dimensions in inches).

```
! material properties (Yeoh)
TB,HYPER,1,3,YEOH    !Activate 3-term Yeoh data table
TBDATA,1,191.322     !Define C1
TBDATA,2,4.9415      !Define C2
TBDATA,3,0.5346      !Define C3
TBDATA,3,5.0E-5      !Define first incompressibility parameter

TB,CDM,1,,3,PSE2     !modified Ogden-Roxburgh Mullins effect
TBDATA,1,2.34,10.0,0.222
```

11.3.3 Loadings and Boundary Conditions

The loading was completed in two steps. In the first step, the rigid plate moved up 0.2 inches to compress the tire (see Figure 11.3a). In the second step, the tire rotated three revolutions under compression (see Figure 11.3b).

11.3.4 Results

Figure 11.4 shows the deformation and stresses of the tire under compression at the end of the first step. The stresses on the tire in the second loading steps, which are very close to each other but lower than those at the end of the first step, are presented in Figure 11.5. The damage in the first revolution is plotted in Figure 11.6, and at the end of the second and third revolutions in Figures 11.7 and 11.8, respectively. Figures 11.9–11.12 illustrate the reaction forces and reaction moments with time and revolutions. Both show that the reaction force and moment have a sharp drop at the end of the first revolution, which is consistent with the damage change.

11.3.5 Discussion

A 3D finite element model of a tire compressed on a rigid road was developed to simulate the Mullins effect of the tire. The variation of the damage variable with the

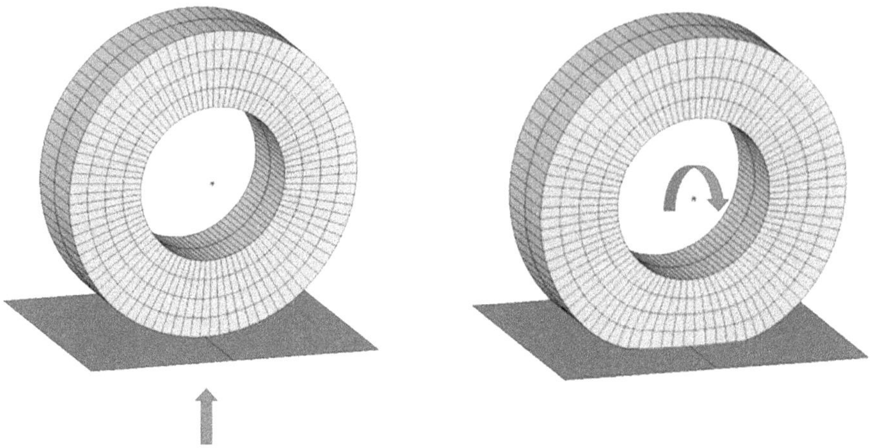

FIGURE 11.3 Loadings. (a) Plate moves up to compress the rubber tire. (b) Rubber tire rotates three rounds.

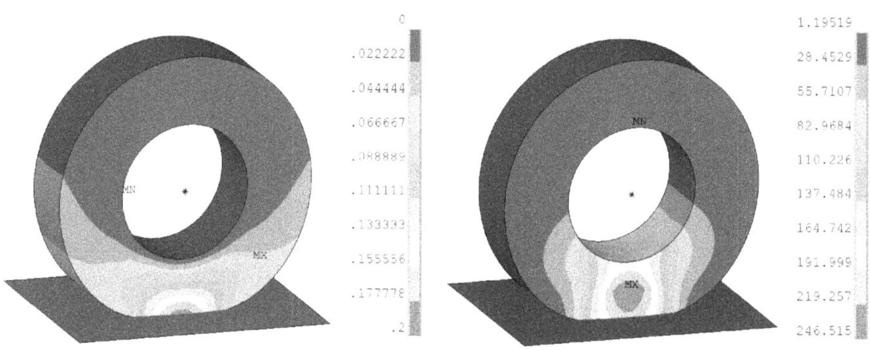

FIGURE 11.4 Tire under compression at the end of the first step.

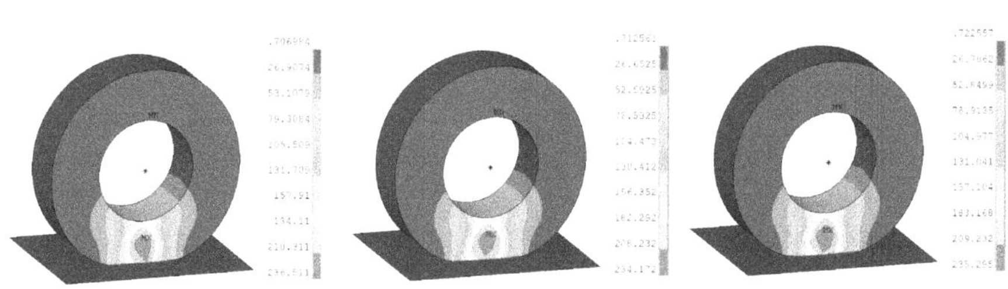

FIGURE 11.5 Stresses of the tire in the second loading steps (Psi). (a) End of first revolution. (b) End of second revolution. (c) End of third revolution.

rotation of the tire is consistent with the reaction force and reaction moment with time and the revolution of the tire.

The damage variable in the Mullins effect differs from other damage models, such as the composite damage in Chapter 19. The damage variable in the Mullins effect is associated with the virgin material strain-energy potential W_0 and the

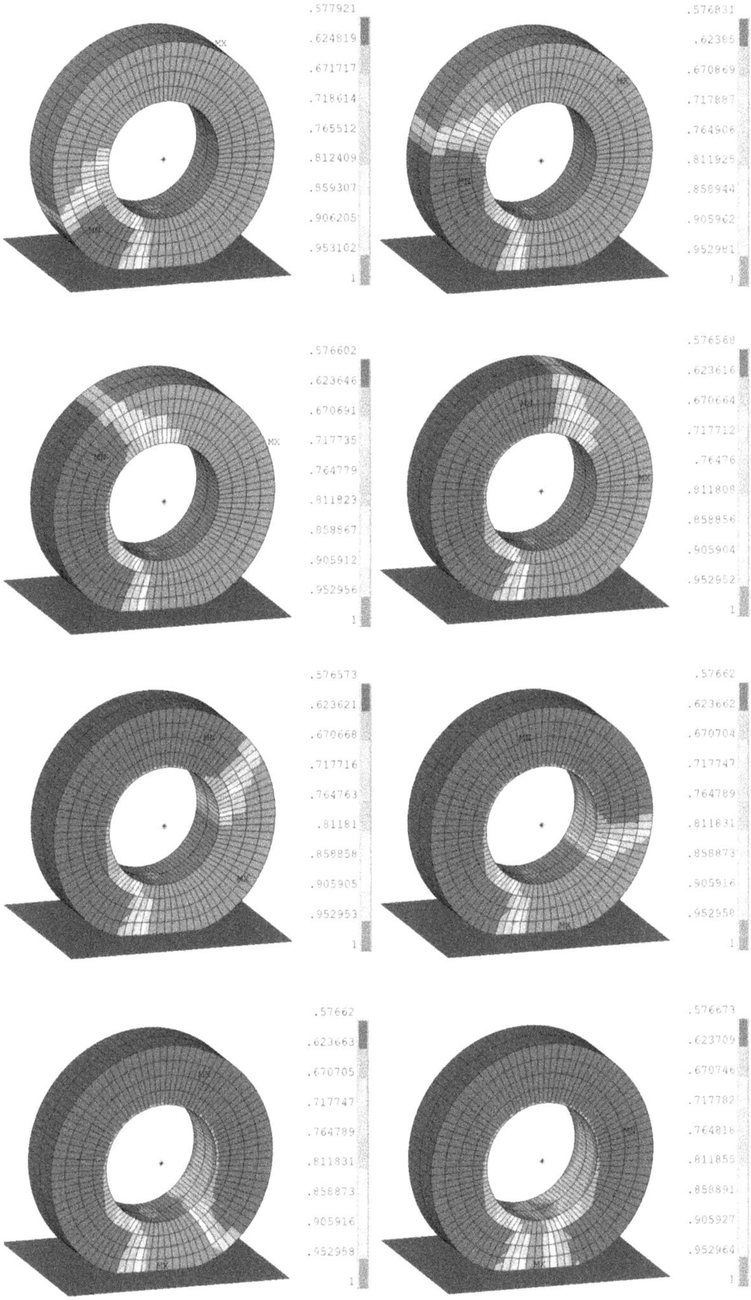

FIGURE 11.6 Damage in the first revolution. (a) Rotate 45°. (b) Rotate 90°. (c) Rotate 135°. (d) Rotate 180°. (f) Rotate 225°. (f) Rotate 270°. (g) Rotate 315°. (h) Rotate 360°.

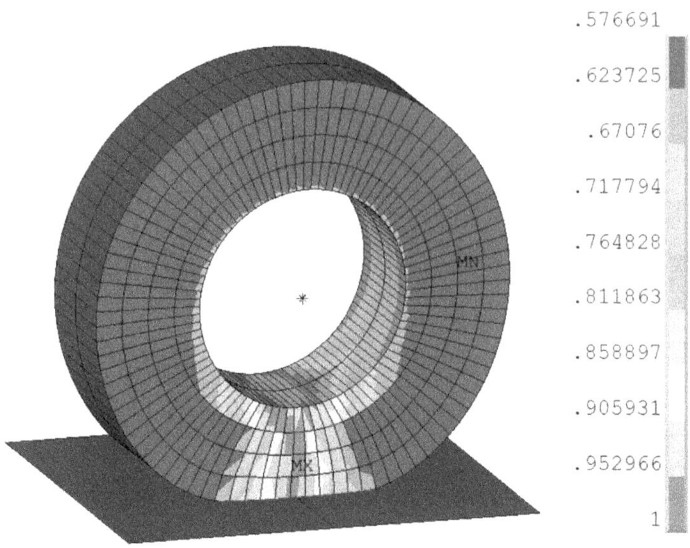

FIGURE 11.7 Damage of the rubber tire at the end of the second revolution.

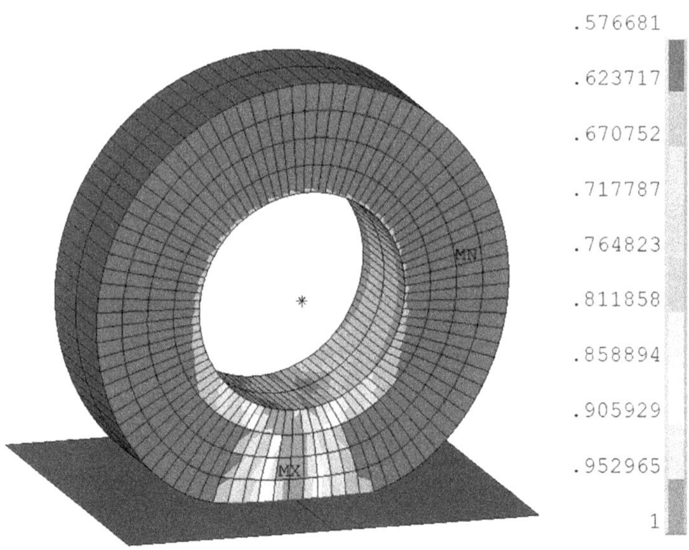

FIGURE 11.8 Damage of the rubber tire at the end of the third revolution.

maximum virgin potential over time W_m. If $W_0 = W_m$, the damage variable becomes 1. If $W_0 < W_m$, the damage variable is less than 1. This explains why the red color referring to 1 in the damage contour disappears gradually in the first revolution. After the first revolution, both W_0 and W_m are stable; that is why the reaction force and moment have a slight change in the second and third revolutions. In the Mullins

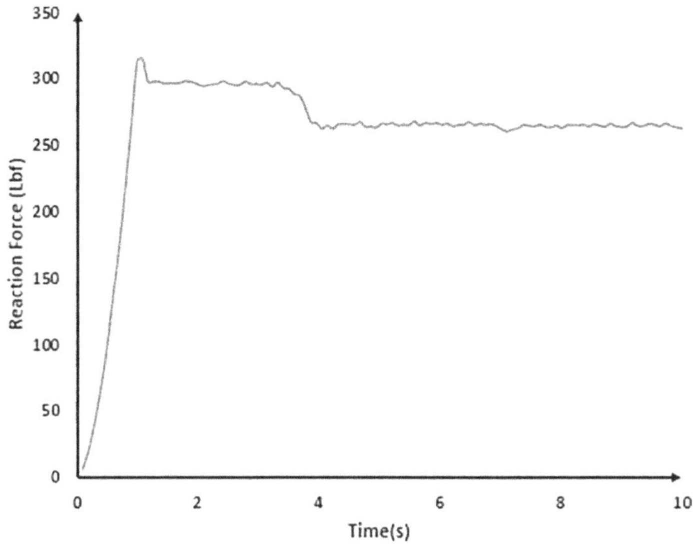

FIGURE 11.9 Reaction force with time.

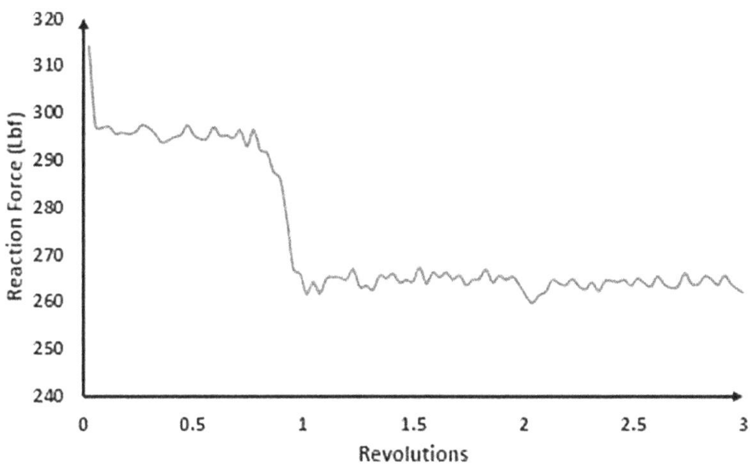

FIGURE 11.10 Reaction force with cycle number.

effect, the physical meaning of the damage variable looks more like the effect variable than the damage variable.

11.3.6 Summary

The Mullins effect of a tire compressed on a rigid road was studied in ANSYS190. The reaction force and moment of the tire match the evolution of the damage variable.

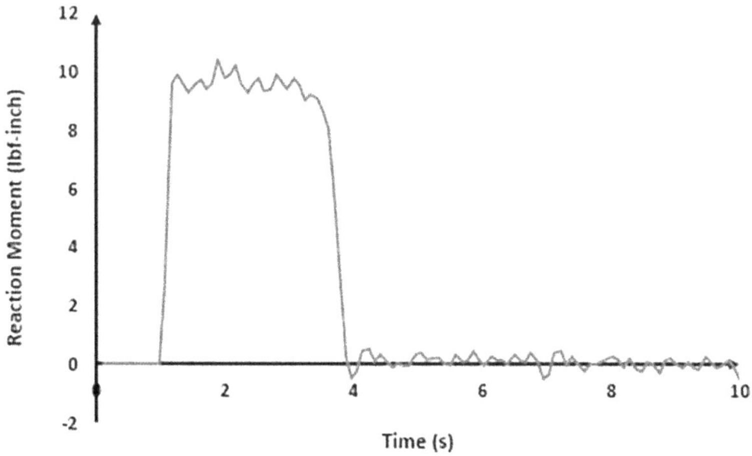

FIGURE 11.11 Reaction moment with time. (a) Deformation. (b) von-Mises stresses (Psi).

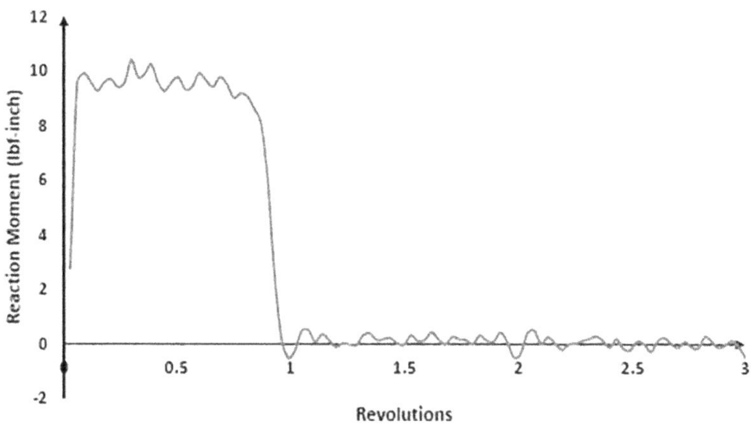

FIGURE 11.12 Reaction moment with cycle number.

REFERENCES

[1] Harwood, J.A.C., Mullins, L., and Payne, A.R., Stress softening in natural rubber vulcanizates. Part II Stress softening in pure gum and filler loaded rubbers, *Journal of Applied Polymer Science*, Vol. 9, 1965, pp. 3011–3021.

[2] Harwood, J.A.C., and Payne, A.R., Stress softening in natural rubber vulanizates, Part III, Carbon black-filled vulanizates, *Journal of Applied Polymer Science*, Vol. 10, 1966, pp. 315–324.

[3] Harwood, J.A.C., and Payne, A.R., Stress softening in natural rubber vulcanizates, Part IV, Unfilled vulcanizates, *Journal of Applied Polymer Science*, Vol. 10, 1966, pp. 1203–1211.

[4] Diani, J., Fayolle, B., and Gilormini, P., A review of Mullins effects, *European Polymer Journal*, Vol. 45, 2009, pp. 601–612.

[5] Bergstrom, J., *Mechanics of Solid Polymers, Elsevier*, Amsterdam, 2015.
[6] Ogden, R.W., and Roxburgh, D.G., A pseudo-elastic model for the Mullins effect in filled rubber, *Proceedings of the Royal Society of London, Series A*, Vol. 455, 1999, pp. 2861–2877.

12 UserHyper for Modeling Hyperelastic Materials

ANSYS provides a powerful tool for customers to develop their own material models. Chapter 6 introduces the UserMat implementation. This chapter discusses UserHyper in ANSYS190 and provides one example to reproduce the Veronda-Westman model.

12.1 INTRODUCTION OF SUBROUTINE USERHYPER

Subroutine UserHyper in ANSYS follows the following format [1]:

```
subroutine UserHyper(
     &   prophy, incomp, nprophy, invar,
     &   potential, pInvDer)
```

The above input and output parameters are explained as follows:

nrophy – number of material parameters,
prophy – array of material parameters,
invar – invariants I_1, I_2 , and J,
potential – strain energy W, and
pInvDer – derivative of the strain-energy potential with respect to I_1, I_2 , and J.

plnvDer includes nine terms: $\frac{\partial W}{\partial I_1}$, $\frac{\partial W}{\partial I_2}$, $\frac{\partial^2 W}{\partial I_1 \partial I_1}$, $\frac{\partial^2 W}{\partial I_1 \partial I_2}$, $\frac{\partial^2 W}{\partial I_2 \partial I_2}$, $\frac{\partial^2 W}{\partial I_1 \partial J}$, $\frac{\partial^2 W}{\partial I_2 \partial J}$, $\frac{\partial W}{\partial J}$, and $\frac{\partial^2 W}{\partial J \partial J}$, consecutively from one to nine.

The key of UserHyper is to calculate and output the nine terms of plnvDer by using three invariants: I_1, I_2 , and J.

12.2 SIMULATION OF VERONDA-WESTMAN MODEL

The Veronda-Westman model represents one of the most widely used material models in biomechanical engineering because its exponential form matches a soft tissue's mechanical response. As an example, the subroutine UserHyper was created to simulate the Veronda-Westman model.

12.2.1 Subroutine UserHyper for Veronda-Westman Model

The strain energy density function of the Veronda-Westman model has the following form [2]:

DOI: 10.1201/9781003436317-14

$$W(I_1) = c_1\left[e^{c_3(I_1-3)} - 1\right] - c_2(I_2 - 3) + \frac{1}{d}\left(\frac{J^2-1}{2} - \ln J\right) \tag{12.1}$$

where c_1, c_2, and c_3 are material constants, and d is the material incompressibility parameter.

Calculating the first and second derivatives of the strain-energy potential with respect to the three invariants leads to the following:

$$\frac{\partial W}{\partial I_1} = c_1 c_3 e^{c_3(I_1-3)} \tag{12.2}$$

$$\frac{\partial W}{\partial I_2} = -c_2 \tag{12.3}$$

$$\frac{\partial^2 W}{\partial I_1 \partial I_1} = c_1 c_3^2 e^{c_3(I_1-3)} \tag{12.4}$$

$$\frac{\partial W}{\partial J} = \frac{1}{d}\left(J - \frac{1}{J}\right) \tag{12.5}$$

$$\frac{\partial^2 W}{\partial J \partial J} = \frac{1}{d}\left(1 + \frac{1}{J^2}\right) \tag{12.6}$$

Because the other derivatives are zero, the above was implemented into the UserHyper subroutine.

```
  Subroutine UserHyper(
&                 prophy, incomp, nprophy, invar,
&                 potential, pInvDer)
#include "impcom.inc"
      DOUBLE PRECISION ZERO, ONE, TWO, THREE, HALF, TOLER
      PARAMETER (ZERO  = 0.d00,
     &           ONE   = 1.0d0,
     &           TWO   = 2.d0,
     &           THREE = 3.d0,
     &           TOLER = 1.0d-12)
c
c --- argument list
c
      INTEGER    nprophy
      DOUBLE PRECISION prophy(*), invar(*),
     &           potential, pInvDer(*)
      LOGICAL    incomp
c
c --- local variables
c
      DOUBLE PRECISION i1, i2, jj, c1, c2, c3, oD1, j1
```

```
c
      i1 = invar(1)
      i2 = invar(2)
      jj = invar(3)
      c1 = prophy(1)
      c2 = prophy(2)
      c3 = prophy(3)
      oD1 = prophy(4)
      potential = ZERO
      pInvDer(1) = ZERO
      pInvDer(2) = ZERO
      pInvDer(3) = ZERO

      potential = c1*(exp(c3*(i1-THREE))-ONE)-c2/
TWO*(i2-THREE)
      pInvDer(1) = c1*c3*exp(c3*(i1-THREE))
      pInvDer(2) = -c2/TWO
      pInvDer(3) = c1*c3*c3*exp(c3*(i1-THREE))

      j1 = ONE / jj
      pInvDer(8) = ZERO
      pInvDer(9) = ZERO
      IF(oD1 .gt. TOLER) THEN
         oD1  = ONE / oD1
         incomp = .FALSE.
         potential = potential + oD1*((jj*jj - ONE)*HALF
- log(jj))
         pInvDer(8) = oD1*(jj - j1)
         pInvDer(9) = oD1*(ONE + j1*j1)
      END IF
c
      RETURN
      END
```

A uniaxial tension was simulated with the following defined material model. The results (see Figure 12.1) indicate that even under very small strains, the stress-strain curve still emerges as strongly nonlinear.

```
! DEFINE MATERIAL BY TB, HYPER,,, ,USER
     TB, HYPER, 1, ,4, USER
     TBDATA, 1, 0.542e-3, 0.542e-2, 21.76, 1e-7
```

12.2.2 Study of Mechanical Response of a Breast with a Tumor

Breast cancer is the second leading cancer in women [3]; therefore, early detection and diagnosis are the most important factors in the treatment of this cancer. Elastography, a very fast process, has been recently developed to detect or classify tumors based on the fact that tissue pathological changes often cause significant stiffness changes. The Veronda-Westman model has been used in the study of elastography because its exponential form is consistent with a soft tissue mechanical response.

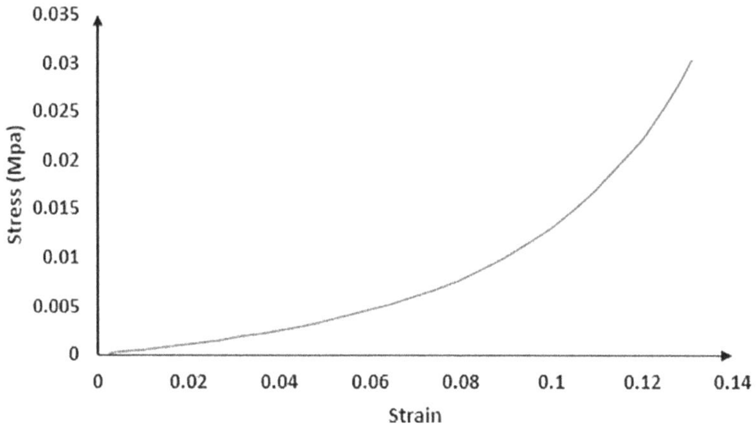

FIGURE 12.1 Stress-strain curve of the Veronda-Westman model.

This study investigated the mechanical response of a breast modeled by the Veronda-Westman model, which is beneficial to the development of elastography.

12.2.2.1 Finite Element Model

A 2D axisymmetrical model was built in ANSYS190 using PLANE182 with keyopt (3)=1 to simulate a breast with a tumor (see Figure 12.2).

12.2.2.2 Material Properties

Apipose, fibroglandular, and tumor tissues were modeled by the Veronda-Westman model with the material parameters in Table 12.1 and defined in ANSYS as follows:

```
TB,HYPER,1,,4,USER
TBDATA,1,2.79E3,1.12E3,1.11,1E-7      ! ADIPOSE TISSUES
MP,DENS,1,0.75E3
TB,HYPER,2,,4,USER
TBDATA,1,5.38E3,2.29E3,2.58,1E-7     !FIBROGLANDULAR TISSUES
MP,DENS,2,1E3
TB,HYPER,3,,4,USER
TBDATA,1,8.20E3,2.15E3,3.39,1E-7     ! TUMOR TISSUES
MP,DENS,3,1E3
```

12.2.2.3 Loadings and Boundary Conditions

The bottom surface was fixed in the vertical direction, and pressure of 1,000 Pa was applied on the surface of the breast to test its mechanical response (see Figure 12.3).

12.2.2.4 Results

The final von-Mises stress plotted in Figure 12.4 indicates that the stresses around the tumor are higher than those around tissues, which demonstrates that elastography is plausible.

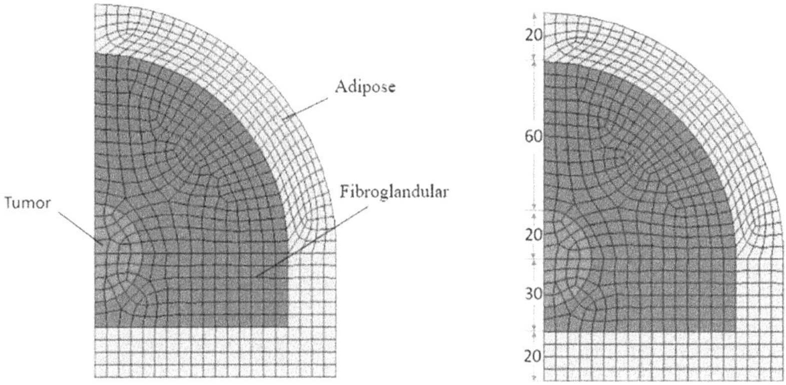

FIGURE 12.2 Finite element model of a breast with a tumor. (a) Finite element model. (b) Geometry (all dimensions in mm).

TABLE 12.1
Material Parameters of Tissues [4]

	Adipose	Fibroglandular	Tumor
c_1(Pa)	2.79e3	5.38e3	8.20e3
c_2(Pa)	1.12e3	2.29e3	2.15e3
c_3	1.11	2.58	3.39
d	1e-7	1e-7	1e-7
Density (kg/m^3)	750	1,000	1,000

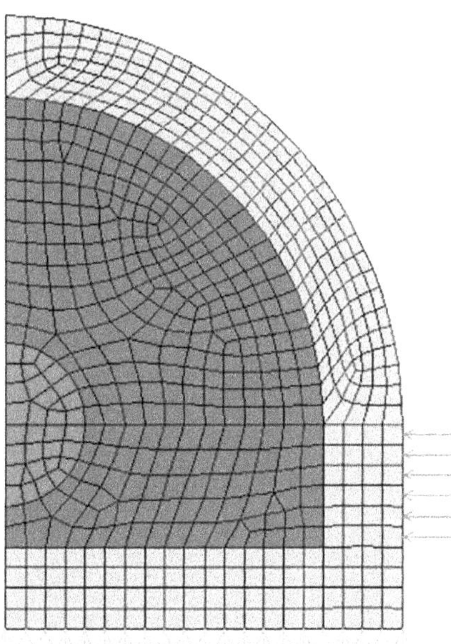

FIGURE 12.3 Boundary conditions.

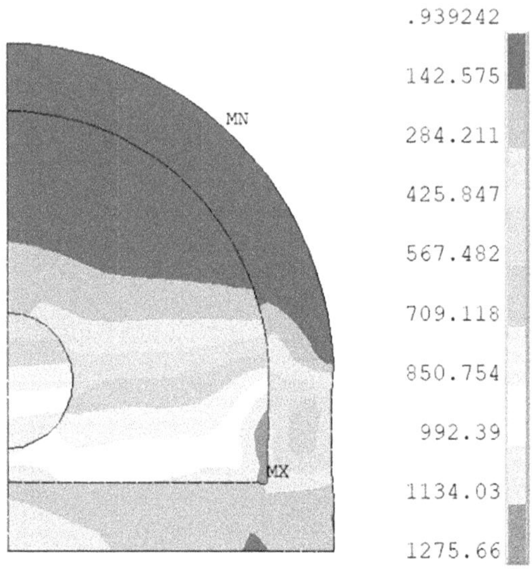

FIGURE 12.4 von-Mises stresses of the breast (Pa).

12.2.2.5 Summary

The Veronda-Westman model was reproduced by subroutine UserHyer with less than thirty lines of code and applied to the study of the mechanical response of a breast with a tumor.

REFERENCES

[1] ANSYS190 help documentation of the help page of product ANSYS190.
[2] Veronda, D.R., and Westmann, R.A., Mechanical characterization of skin-finite deformations, *Journal of Biomechanics*, Vol. 3, 1970, pp. 111–122.
[3] World Health Organization International Agency for Research on Cancer, "World Cancer Report 2008".
[4] Amooshahi, M., and Samani, A., A fast breast nonlinear elastography reconstruction technique using the Veronda-Westman model, *5th Canadian Student Conference on Biomedical Computing and Engineering*, Waterloo, May, 2010.

Part III

Soil

When we construct a big building such as a skyscraper, we must analyze the foundation to ensure the building is stable after its completion. Similarly, when we do a tunnel excavation, we also need to engage in structural analysis. Works like these can benefit from finite element analysis, which involves various material models of the soil. This is the topic of Part III.

Chapter 13 presents a brief introduction to soil. Then, Chapters 14–17 focus on four major material models of soil and their applications, including the Cam Clay model, Drucker-Prager model, Mohr-Coulomb model, and Jointed Rock model.

Soil consists of a solid phase and water. When the soil is under compression, it expels water. That is the soil consolidation discussed in Chapter 18.

This part does not discuss the temperature and time effects on the soil material models because they can be handled in the same way as in Part I.

DOI: 10.1201/9781003436317-15

13 Soil Introduction

Chapter 13 primarily highlights soil structure and the parameters that characterize the soil.

13.1 SOIL STRUCTURE

Sand, silt, and clay are the three main mineral components of soil. Their relative proportions control a soil's texture (see Figure 13.1), and the soil's texture influences the properties of the soil such as porosity, permeability, and water-holding capacity. Sand and silt come from the physical and chemical weathering of the parent rock [1], while clay is produced by the precipitation of the dissolved parent rock. Sand is the least active; clay is the most active, followed by silt. Sand helps the soil resist compaction and increase soil porosity. Silt has a higher specific surface area than sand, which makes silt more chemically and physically active than sand. Clay, due to its very specific surface area and large number of negative charges, enables the soil to retain water and nutrients. From the particle size perspective, sand consists mainly of quart particles with a diameter that ranges from 2.0 to 0.04 mm, silt ranges from 0.05 to 0.002 mm, and clay is as small as 0.002 mm. Soil with a size over 2.0 mm is classified as rock and gravel.

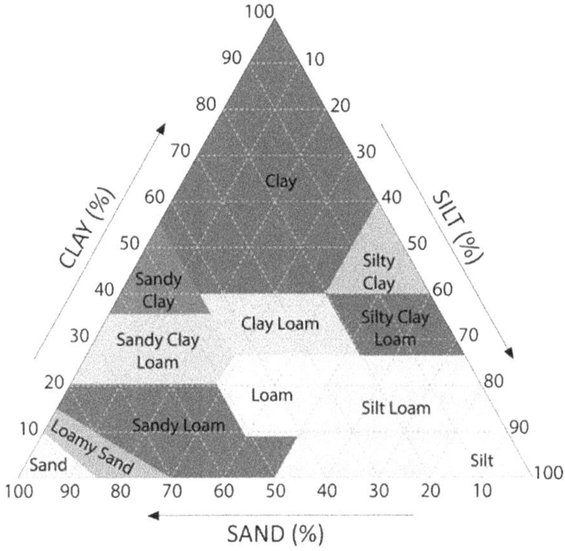

FIGURE 13.1 Soil type. (Kanda Euatham © 123rf.com.)

DOI: 10.1201/9781003436317-16

13.2 SOIL PARAMETERS

Soil is characterized by some parameters, including the following:

1. Density. Soil particle density ranges from 2.60 to 2.75 g/cm^3. The soil bulk density is defined as $\frac{dry\ mass\ of\ soil}{volume\ of\ soil}$, which is always smaller than the soil particle density. Thus, the soil bulk density can be used to characterize soil compaction.
2. Porosity. Soils are composed of solid particles. Voids exist between the particles, in which air, water, or both stay. Porosity is calculated by $1-\frac{bulk\ density}{particle\ density}$.
3. Consistency. Consistency refers to the cohesion and adhesion abilities of soil and its ability to sustain deformation and rupture. It is very useful for predicting cultivation problems and the engineering of foundations [2]. In civil engineering, soil consistency is used prior to construction to evaluate the soil that supports buildings and roads.

REFERENCES

[1] Jenny, H., *Factors of Soil Formation: A System of Quantitative Pedology*, McGraw-Hill, New York, 1941.
[2] Merritt, F.S, and Rickett, J.T., *Building Design and Construction Handbook*, (6th ed.), McGraw-Hill Professional, New York, 2000.

14 Cam Clay Model

Chapter 14 focuses on the Cam Clay model. Sections 14.1 and 14.2 introduce the Modified Cam Clay model, including its advantages and how to define the Modified Cam Clay model in ANSYS. In addition, soil excavation and subsidence of a tower are simulated in Sections 14.3 and 14.4, respectively.

14.1 INTRODUCTION OF MODIFIED CAM CLAY MODEL

The Cam Clay model and Modified Cam Clay model were presented first by the researchers at Cambridge University to describe the behaviors of soft soils [1,2]. The Modified Cam Clay model was developed from the Cam Clay model and has been widely used in geotechnical engineering since predication of soil deformation is crucial in many geotechnical engineering problems. Thus, the following introduction is limited to the Modified Cam Clay.

In the Modified Cam Clay model, the voids between the solid particles are assumed to be full of only water. When the soil is loaded, the water is expelled from the voids, and the volume of the soil changes. The Modified Cam Clay yield function is an elliptic function [3]

$$\frac{q^2}{p'^2} + M^2\left(1 - \frac{p'_c}{p'}\right) = 0 \tag{14.1}$$

In the p'-q plane, the Modified Cam Clay yield function is an ellipse (see Figure 14.1).

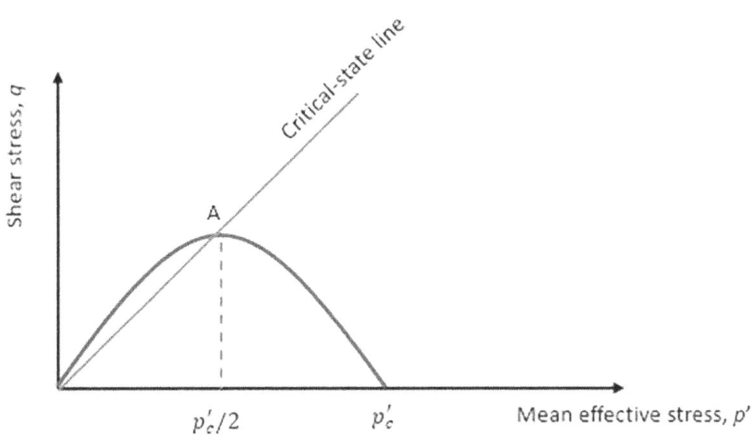

FIGURE 14.1 Yield function of the Modified Cam Clay model.

DOI: 10.1201/9781003436317-17

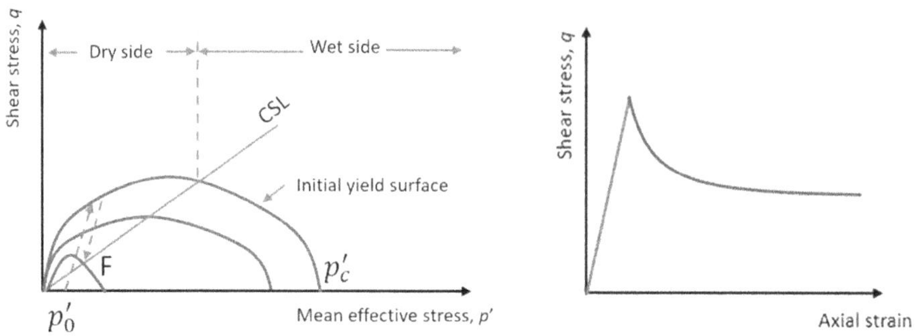

FIGURE 14.2 Softening behavior of the Modified Cam Clay model [3]. (a) Development of a yield surface (dash line). (b) Stress-strain curve.

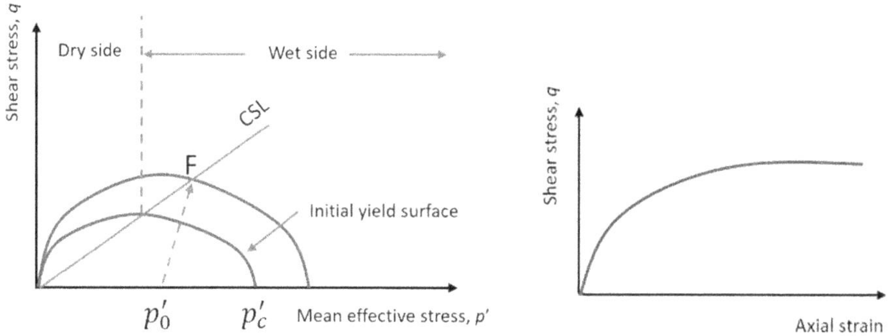

FIGURE 14.3 Hardening behavior of the Modified Cam Clay model [3]. (a) Development of a yield surface (dash line). (b) Stress-strain curve.

The Modified Cam Clay model can predict unlimited soil deformations (including softening and hardening) with a constant stress or volume when a critical state is reached. Figure 14.2 illustrates the softening behavior of the Modified Cam Clay model. Before the effective stress path (dash line) touches the initial yield surface, the heavily overconsolidated soil is elastic. After the effective stress path touches the initial yield surface, the soil starts to contract (soften); it undergoes plasticity until the effective stress path reaches the F point, where the failure occurs. Figure 14.3 depicts the hardening behavior of the Modified Cam Clay model. Within the initial yield surface, the normally consolidated and lightly overconsolidated soil is elastic. After the effective stress path reaches the initial yield surface, the soil has plastic strains. The yield surface becomes larger (hardening) until the effective stress path reaches point F, where the soil fails.

Another advantage of the Modified Cam Clay model is that the yield surface (see Figure 14.1) is very smooth, which makes the solution converge easily. As a result, the Modified Cam Clay has wide applications in geotechnical engineering.

14.2 MODIFIED CAM CLAY MODEL IN ANSYS

The Modified Cam Clay model in ANSYS consists of the elastic component and plastic component, which are individually introduced.

14.2.1 Elastic Component

The stiffness matrix **C** is composed of the bulk modulus and shear modulus [4],

$$\mathbf{C} = k_b \mathbf{I} \otimes \mathbf{I} + 2\mu \mathbf{I}_{dev} \tag{14.2}$$

where k_b – bulk modulus;

μ – shear modulus;

I – the second-order identity tensor; and

$\mathbf{I}_{dev}$ – the fourth-order deviatoric projection tensor.

During the compression of the soil, the bulk modulus is a function of the elastic volumetric strain.

$$k_b = \left(p + p_t^{el}\right)\left(\frac{1+e_0}{\kappa} e^{\varepsilon_{vol}^{el}}\right) \tag{14.3}$$

where κ is the swell index, p_t^{el} is the elastic limit of tensile strength, and e_0 is the initial void ratio.

With a constant Poisson's ratio, the shear modulus is expressed by

$$\mu = \frac{3k_b(1-2v)}{2(1+v)} \tag{14.4}$$

where v is Poisson's ratio.

Since the bulk modulus is a function of the current elastic volumetric strain, and the elastic volumetric strain increases with the loading, the bulk modulus and the shear modulus also increase with the loading. Therefore, in the elastic stage of the soil, the constitutive law of the soil is strongly nonlinear.

In the definition of the porous elasticity model in ANSYS, these material parameters, including swell index κ, elastic limit of tensile strength p_t^{el}, Poisson's ratio v, and initial void ratio e_0, are defined using the following format [4]

```
TB, PELAS,,,,POISSON
TBDATA, 1, κ, pt^el ,v,e0
```

Furthermore, if the initial stress p0 exists in the soil, it can be specified as [4]

```
INISTATE, SET, DTYP, STRE
INISTATE, DEFI,ALL,,,,-P0,-P0,-P0   ! NEGATIVE MEANS
COMPRESSION
```

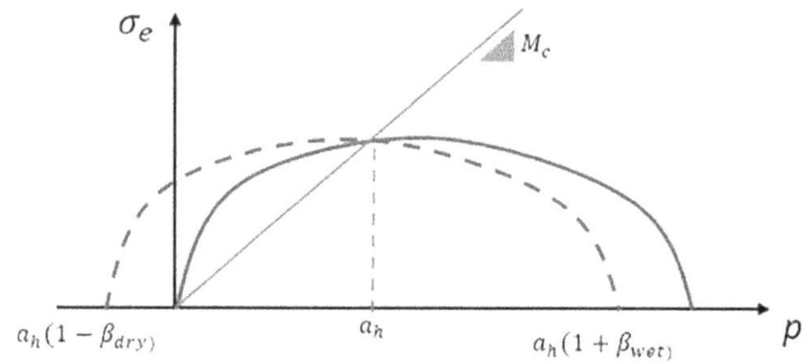

FIGURE 14.4 Yield function of the Extended Cam Clay model [4].

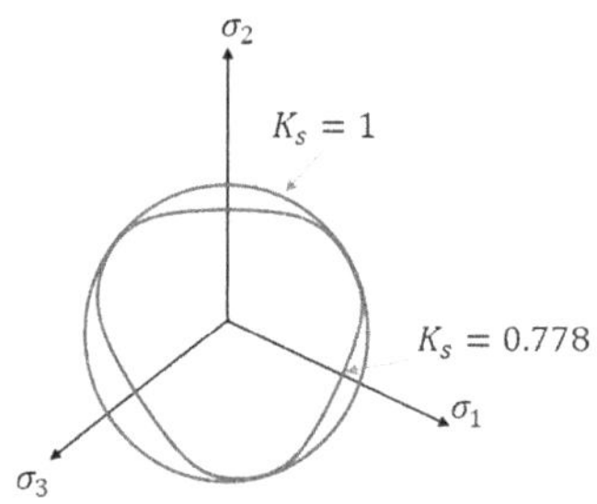

FIGURE 14.5 Various yield functions with different K_s.

14.2.2 Plastic Component

In ANSYS, the plastic part of the Modified Cam Clay model was developed based on the Extended Cam Clay model with the following yield function (see Figure 14.4) [4]:

$$f_{cc} = \frac{1}{\beta^2}\left(\frac{p}{a_h} - 1\right)^2 + \left(\frac{t}{M_c a_h}\right)^2 - 1 \tag{14.5}$$

in which β is used to modify the shape of the yield surface, including β_{dry} for the dry side and β_{wet} for the wet side. a_h is the parameter that controls the size of the yield surface. M_c is the slope of the critical state line. Parameter t is defined by

$$t = \frac{\sigma_e}{g} = \frac{\sigma_e^3(1+K_s)+13.5(1-K_s)J_3}{2K_s\sigma_e^2} \tag{14.6}$$

In the above equation, σ_e is the von-Mises effective stress, J_3 is the third principal invariant of the stress tensor, and K_s modifies the shape of the yield surface (see Figure 14.5). The relation between the void ratio and the pressure during plastic deformation is given as

$$de = -\lambda_F d(\ln p) \tag{14.7}$$

Thus, seven parameters are defined in the Modified Cam Clay model in ANSYS [4]:

```
TB, SOIL,1,,,CAMCLAY
TBDATA,1, λ_F, M_c a_h0, a_h,min, β_dry
TBDATA,6, β_wet, K_s
```

14.3 SIMULATION OF SOIL EXCAVATION

Soil excavation, which was simulated using the Modified Cam Clay model as follows, is a common activity in the field of civil engineering.

A two-dimensional model with a width of 5 m and a height of 6 m was built using PLANE182 with the plane strain option (keyopt(3)=2). The material parameters of the Cam Clay model were selected as follows [5]:

$$\kappa = 0.054, \ M_c = 1.4, \ \lambda_F = 0.37, \ E_0 = 2880\text{Pa}, \ v = 0.35$$

The bottom was fixed in the vertical direction, while the two walls were fixed in the horizontal direction (see Figure 14.6).

The loading involves two steps. The self-weight was loaded in the first step. Then a 3 m×4 m area was excavated by the EKILL command.

Figures 14.7 and 14.8 illustrate the von-Mises stresses and plastic strains after excavation, which confirm that excavation does change the distribution of stresses and plastic strains.

FIGURE 14.6 Excavation model.

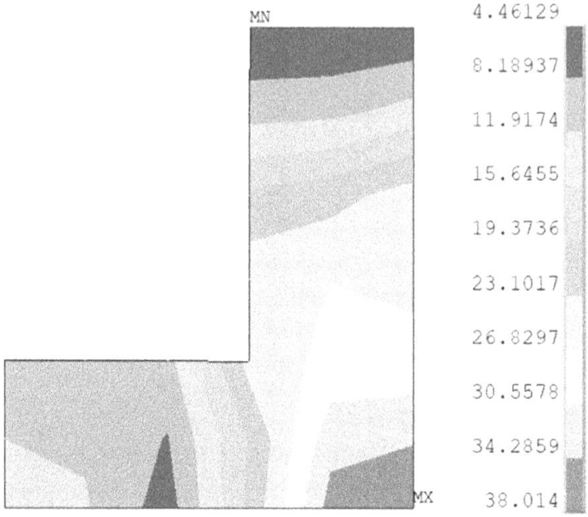

FIGURE 14.7 von-Mises stresses of the soil after excavation (Pa).

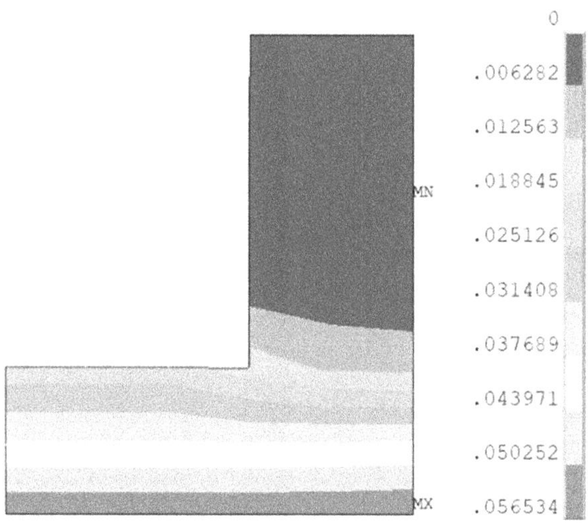

FIGURE 14.8 von-Mises plastic strains of the soil after excavation.

14.4 SIMULATION OF A TOWER ON THE GROUND BY MODIFIED CAM CLAY MODEL

Subsidence occurs due to the deformation of the soil when a building is constructed. Since subsidence is always associated with the stability of a building, it is very important in the structural design [3]. In this study, the Modified Cam Clay model was applied to study the subsidence of the tower built, which provides an example for a similar structural analysis.

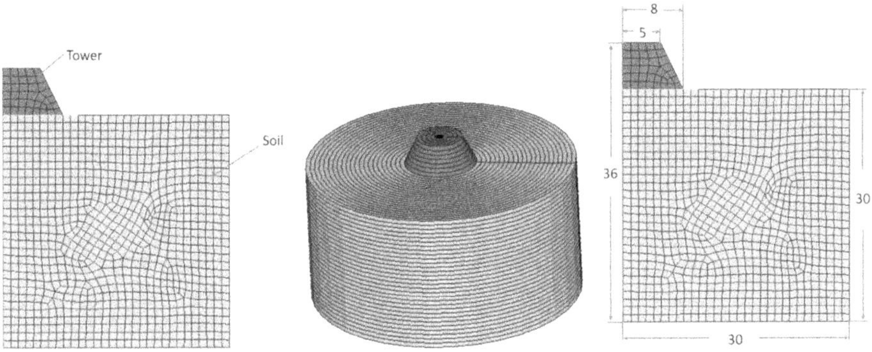

FIGURE 14.9 Finite element model of a tower on the ground. (a) 2D axisymmetrical model. (b) 3D model. (c) Geometry (all dimensions in meter).

14.4.1 Finite Element Model

Towers are generally conical. Thus, one cross-section of the model was created in ANSYS190 (see Figure 14.9), which was meshed by PLANE182 with an axisymmetrical keyoption (keyopt(3)=1).

14.4.2 Material Properties

The tower was modeled with a linear elastic material with Young's modulus 2e10 Pa and density 5,000 kg/m^3. The soil was simulated with porous elasticity and Modified Cam Clay model. The corresponding APDL commands to define the soil material parameters are listed as follows:

```
TB,PELAS,1,,,POISSON
TBDATA,1,0.026,5.0e3,0.28, 0.889
TB,SOIL,1,,,CAMCLAY
TBDATA,1,0.1174,1,103e3, 103e3,1
TBDATA,6,1,1
```

14.4.3 Contact Definition

The contact between the tower and the ground was defined as standard contact. Because the tower is much stiffer than the ground, the contact areas connected with the tower and the ground were modeled with TARGE169 and CONTA172, respectively (see Figure 14.10).

14.4.4 Loadings and Boundary Conditions

The whole model was loaded with self-weight using the command ACEL, and the ground was constrained on the boundary (see Figure 14.11).

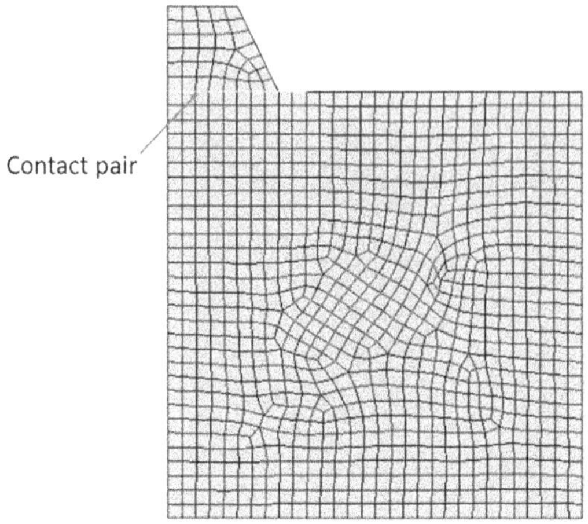

FIGURE 14.10 Contact in the model.

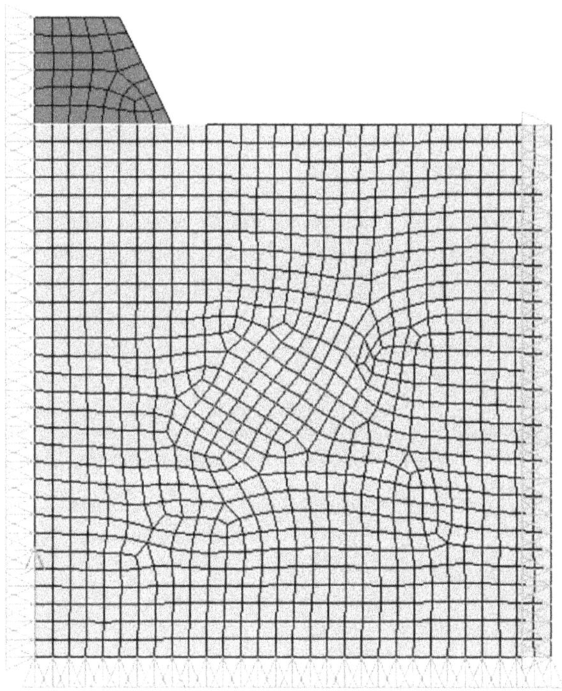

FIGURE 14.11 Boundary conditions and loadings.

14.4.5 Results

The deformation of the whole model is illustrated in Figure 14.12, which indicates the occurrence of a 2.69 m vertical displacement. The von-Mises stresses and the plastic strains of the soil are plotted in Figures 14.13 and 14.14, respectively.

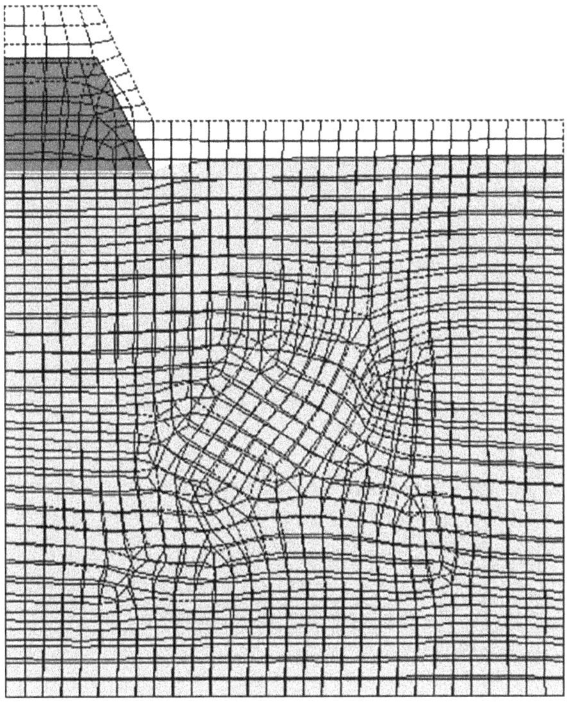

FIGURE 14.12 Deformation (Dmax = 2.69 m).

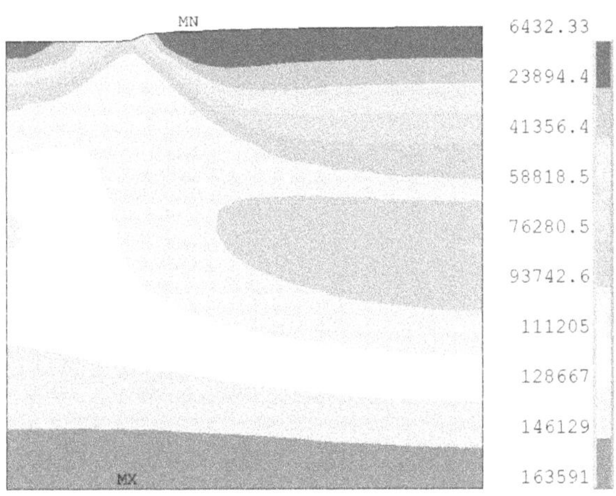

FIGURE 14.13 von-Mises stresses of the soil (Pa).

Obviously, the stresses increase with the depth of the soil because of the weight. The distribution of the plastic strains looks like that of the stresses, except that the maximum plastic strain of 0.075 occurs at the contact edge between the ground and the tower.

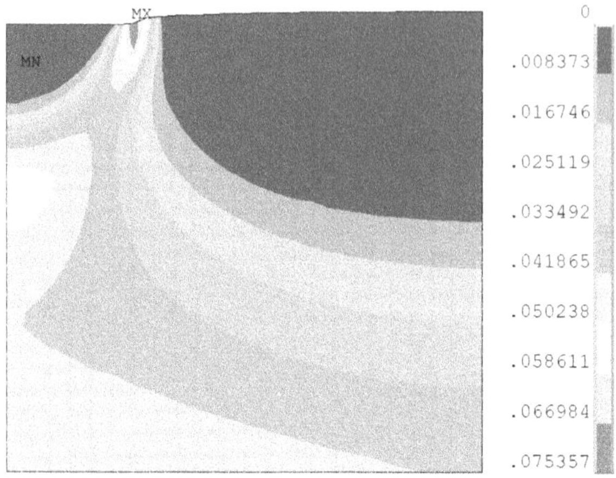

FIGURE 14.14 von-Mises plastic strains of the soil.

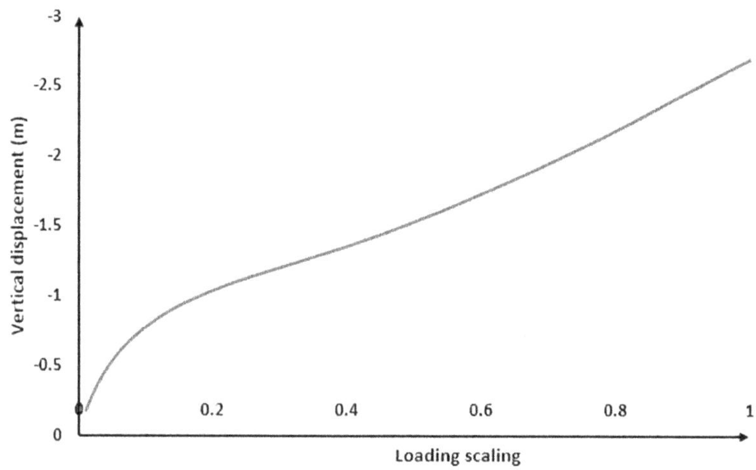

FIGURE 14.15 Vertical displacement with loading scaling.

The history of the vertical displacement of the tower with the loading is presented in Figure 14.15, which shows that, at the beginning stage, the vertical displacement increased quickly with the increase in loading. After 10% of the loading was completed, the vertical displacement increased much more slowly than at the beginning stage.

14.4.6 Discussion

A 2D axisymmetrical finite element model was constructed to simulate the subsidence of the tower built using the porous elasticity and Modified Cam Clay models. The large vertical displacement of the tower occurs after the whole model is loaded with self-weight, which confirms the subsidence of the tower.

The loading history of the vertical displacement of the tower reveals the features of the soil model. The stiffness of the soil at the elasticity increases with the pressure that is linked to the loading. Initially, the stiffness of the soil is very small; therefore, the vertical displacement changes rapidly. With the increase in loading, however, the pressure increases as well, which increases the stiffness of the soil. Thus, the vertical displacement changes gradually slow down. That is why the vertical displacement changes with the loading differ before and after 10% of the loading.

14.4.7 Summary

The subsidence of the tower was simulated in ANSYS190 using the porous elasticity and Modified Cam Clay models. The computational results are consistent with the features of the material models.

REFERENCES

[1] Roscoe, K.H., Schofield, A.N., and Wroth, C.P., On the yielding of soils, *Geotechnique*, Vol. 8, 1958, pp. 22–53.
[2] Roscoe, K.H., and Burland, J.B., On the generalised stress-strain behaviour of 'wet' clay, *Engineering Plasticity*, pp. 535–609, Cambridge University Press, Cambridge, 1968.
[3] Helwany, S., *Applied Soil Mechanics with ABAQUS Application*, John Wiley & Sons, Inc., Hoboken, 2007.
[4] ANSYS190 Help Documentation in the help page of product ANSYS190.
[5] Borja, R.I., Cam-Clay plasticity, Part II: Implicit integration of constitutive equation based on a nonlinear elastic stress predictor, *Computer Methods in Applied Mechanics and Engineering*, Vol. 88, 1991, pp. 225–250.

15 Drucker-Prager Model

In addition to the Cam Clay model, the Drucker-Prager model is also widely used in geomechanics. Chapter 15 briefly introduces the Drucker-Prager model. Section 15.1 defines it in ANSYS and describes its application, while Sections 15.2 and 15.3 focus on its application for the concrete slump test and the analysis of a soil-arch interaction, respectively.

15.1 INTRODUCTION OF DRUCKER-PRAGER MODEL

The Drucker-Prager model is a pressure-dependent model that is widely applied for the study of granular (frictional) materials such as soils, rocks, and concrete. In ANSYS, some material models related to the Drucker-Prager model are available, including the classic Drucker-Prager model, Extended Drucker-Prager (EDP) model, Extended Drucker-Prager Cap model, and Drucker-Prager Concrete model. These models all follow the same yield function [1]:

$$f(\sigma,\sigma_y)=\sigma_e+\alpha\frac{1}{3}tr(\sigma)-\sigma_y \tag{15.1}$$

where σ_e is the equivalent stress, σ_y is the uniaxial yield stress, and α is the pressure sensitivity.

α is a function of internal frictional angle ϕ,

$$\alpha=\frac{6\sin\phi}{\sqrt{3}(3-\sin\phi)} \tag{15.2}$$

and σ_y is expressed in terms of a function of ϕ and cohesion value c

$$\sigma_y=\frac{6c\cdot\cos\phi}{\sqrt{3}(3-\sin\phi)} \tag{15.3}$$

The yield function of the Drucker-Prager model in π plane is circular (see Figure 15.1). Because the yield surface of the Drucker-Prager model is smooth, it poses no convergence problem in engineering. Thus, the Drucker-Prager model is often used for the simulation of rocks and concrete.

In ANSYS, the Extended Drucker-Prager (EDP) model is specified by

```
! define the yield surface
TB, EDP,,,,LYFUN
TBDATA,1, α, σy
! define the plastic flow potential
TB,EDP,,,,LFPOT
TBDATA,1,ᾱ   !ᾱ is pressure sensitivity
```

DOI: 10.1201/9781003436317-18

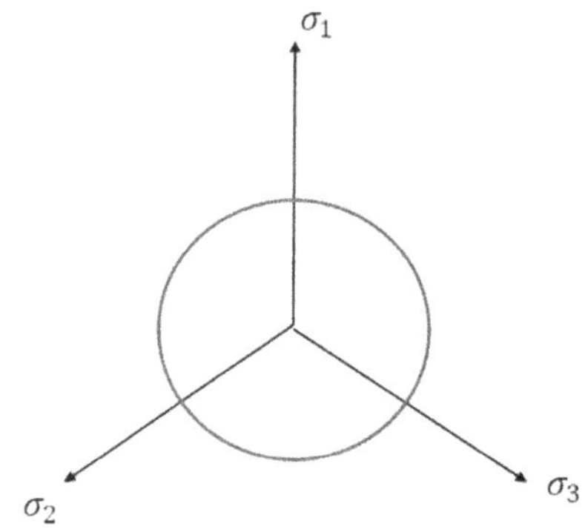

FIGURE 15.1 Yield surface of Drucker-Prager model in π plane.

The definition of other Drucker-Prager models can also be found in the ANSYS190 Help Documentation [1].

15.2 SIMULATION OF CONCRETE SLUMP TEST

The slump test is conducted on fresh, wet concrete to evaluate its consistency and ability to flow. After a conical mold is filled with concrete, the mold is removed. The concrete is deformed under its own weight. The slump test as a standard test was simulated using the Extended Drucker-Prager model.

A two-dimensional axisymmetrical model was created to simulate the concrete in a conical mold (see Figure 15.2).

The material parameters for the concrete were selected as [2]: $\sigma_y = 1.844e - 3$, $\beta = 0.581$

The material data come from the ABAQUS manual without units [2].

The bottom of the model was fixed in the vertical direction, and the whole concrete was loaded with self-weight ($g = 0.666$).

After the solution was quickly completed, the plastic strains in the vertical direction were plotted in Figure 15.3, which shows the values growing larger when closer to the bottom.

15.3 STUDY OF A SOIL-ARCH INTERACTION

Zienkiewicz analyzed a strip footing on soil [3]. This study analyzed this problem by using the Extended Drucker -Prager model in ANSYS190 and compared the model's convergence pattern with that of the Mohr-Coulomb model.

15.3.1 Finite Element Model

Figure 15.4 [4] illustrates a simple soil-arch interaction model. The arch has a 3 m span, a 750 mm rise, and a 215 mm thickness. The arch and soil were meshed by Plane182 with a plane strain state. The contact between the soil and arch barrel interface was

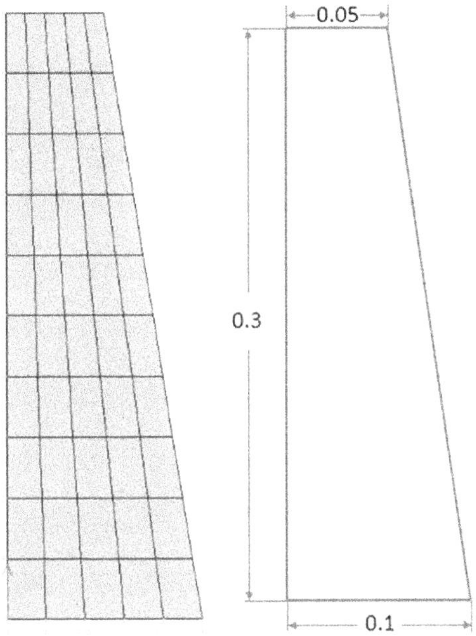

FIGURE 15.2 Finite element model of the slump test. (a) Finite element model. (b) Geometry.

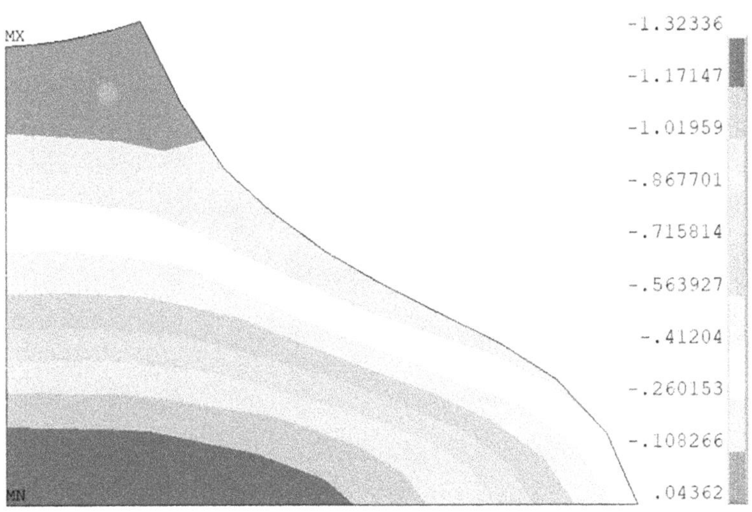

FIGURE 15.3 Plastic strains in the vertical direction.

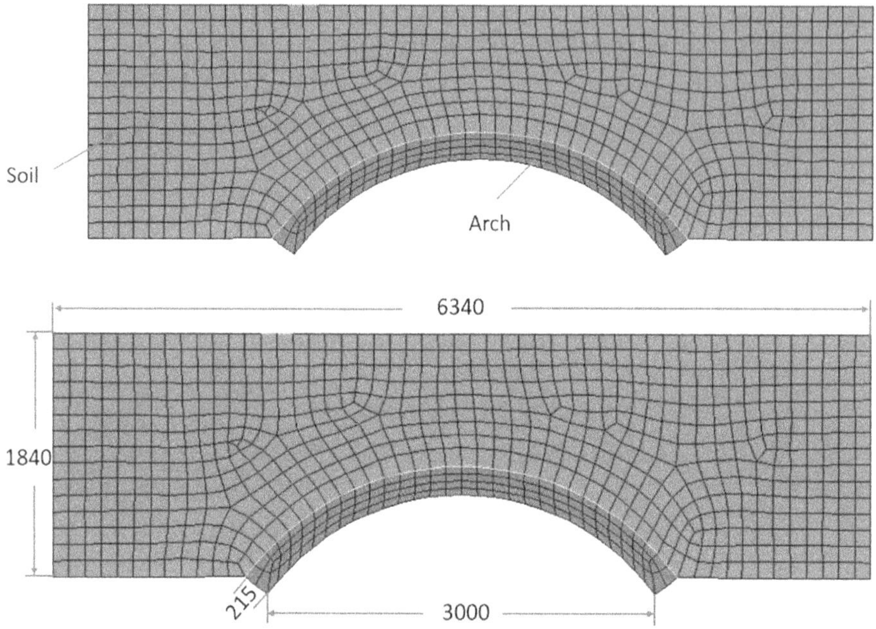

FIGURE 15.4 Finite element model of a soil-arch interaction. (a) Finite element model. (b) Geometry (all dimensions in mm).

assumed to be standard contact and meshed by CONTA173 and TARGE170. The entire model consists of 818 elements and 856 nodes.

15.3.2 Material Properties

The arch barrel top was regarded as elastic with a Young's modulus of 16 GPa and a Poisson ratio of 0.2. The soil was modeled as Extended Drucker-Prager material [4]:

```
! set nonlinear parameters for intact rock
beta=1.05          ! Drucker-Prager parameter (slope of cone)
sigma_y=0.0224     ! Drucker-Prager parameter (strength)
TB,EDP,3,1,2,LYFUN
tbdata,1,beta,sigma_y

TB,EDP,3,1,2,LFPOT
tbdata,1,beta
```

The coefficient of friction between the soil and the arch barrel was assumed to be 0.7.

15.3.3 Boundary Conditions and Loadings

The whole soil area was constrained in the normal directions of the boundary (see Figure 15.5), and the two ends of the arch were constrained with all degrees of freedom.

A pressure of 0.3 MPa was applied to a part of the top edge of the soil.

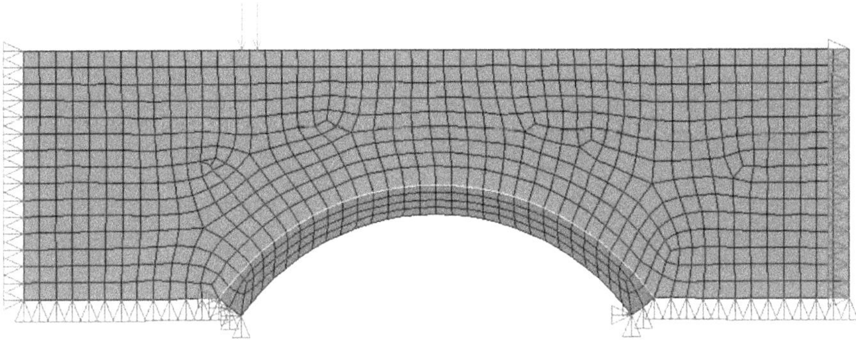

FIGURE 15.5 Boundary conditions and loadings.

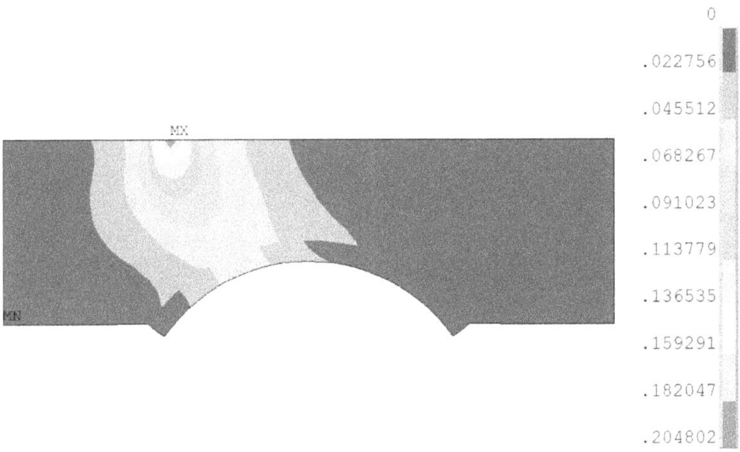

FIGURE 15.6 Deformation of the model (mm).

15.3.4 Results

The final deformation, von-Mises stresses, and plastic strains of the whole model are presented in Figures 15.6–15.8, respectively. The maximum displacement of 0.2 mm occurs at the loading area, and the maximum von-Mises stresses of 0.48 MPa appear within the arch because the arch is 16 times stiffer than the soil. The maximum plastic strain of 1.2e-4 is just a little below the loading area.

Figure 15.9 plots the convergence pattern of the model during the solution, which indicates it converges quickly and finishes all computations in less than 30 iterations.

15.3.5 Discussion

A 2D finite element model of a soil-arch interaction was developed, in which the soil was modeled by the Extended Drucker-Prager model. The stress state and plastic strain distribution were obtained after the solution was complete.

Compared to the Mohr-Coulomb model in Chapter 16, the Drucker-Prager model converges quickly because of its smooth yield surface, which is used in the standard

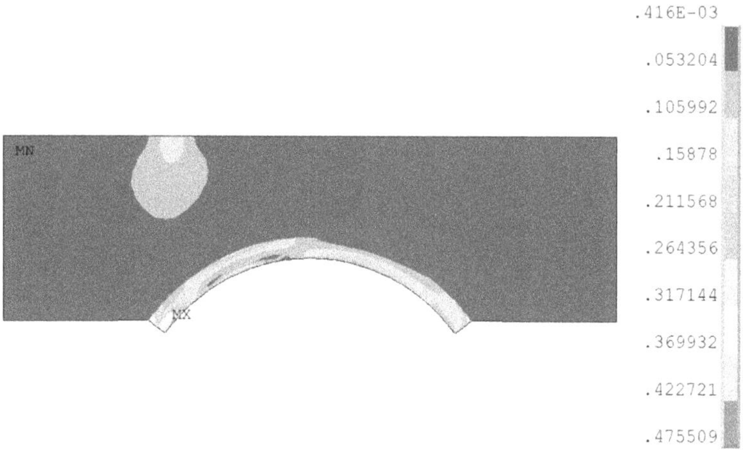

FIGURE 15.7 von-Mises stresses of the model (MPa).

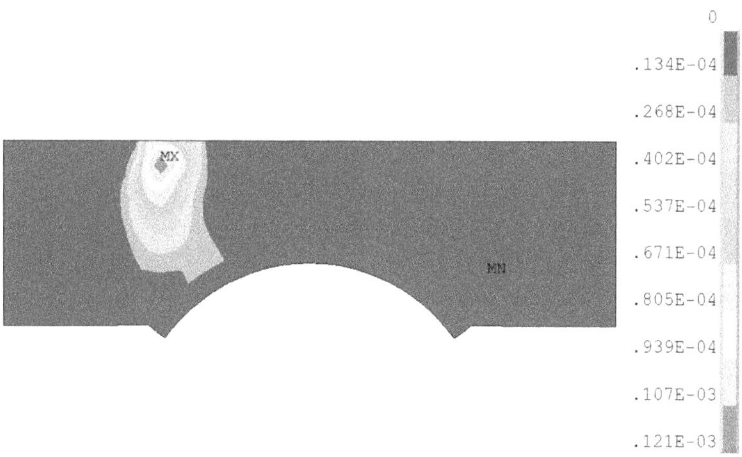

FIGURE 15.8 von-Mises plastic strains of the model.

Newton-Raphson method to solve the nonlinear equations. On the other hand, the yield surface of the Mohr-Coulomb model has sharp angles. Therefore, the cutting-plane algorithm is used to solve the nonlinear equations in the Mohr-Coulomb models, which requires many iterations to complete.

15.3.6 Summary

The soil-arch interaction was simulated in ANSYS190 by the Extended Drucker-Prager model. It converges much more quickly than the Mohr-Coulomb model.

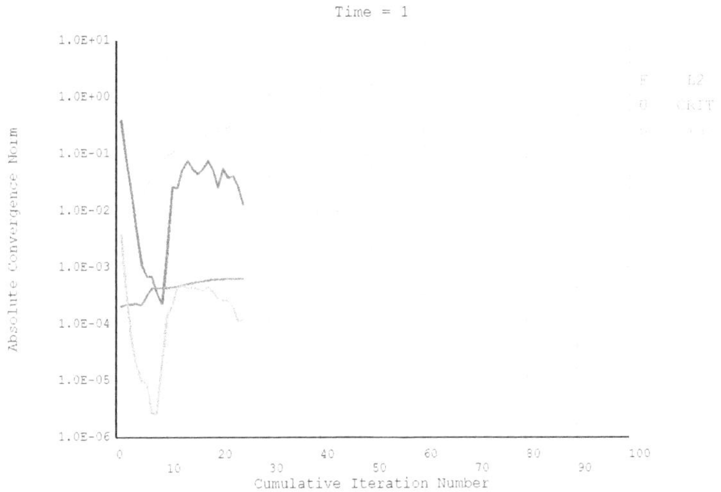

FIGURE 15.9 Convergence pattern of the model.

REFERENCES

[1] ANSYS190 Help Documentation in the help page of product ANSYS190.

[2] ABAQUS version 6.6 Documentation.

[3] Zienkiewicz, O.C., Humpheson, C., and Lewis, R.W., Associated and non-associated viscoplasticity and plasticity in soil mechanics, *Geotechnique*, Vol. 25, 1975, pp. 671–689.

[4] Wang, J., and Melbourne, C., Finite element analyses of soil-structure interaction in masonry arch bridges, *Arch's 07 - 5th International Conference on Arch Bridge*, Madeira, September 2007, pp. 515–523.

16 Mohr-Coulomb Model

Chapter 16 primarily discusses the Mohr-Coulomb model. The introduction of the Mohr-Coulomb model, definition of the Mohr-Coulomb model in ANSYS, and application of the Mohr-Coulomb model for the concrete slump test and slope stability analysis are presented in Sections 16.1–16.4, respectively.

16.1 INTRODUCTION OF MOHR-COULOMB MODEL

When the soil is under triaxial compression (see Figure 16.1), it fails along a failure plane at an angle θ with respect to σ_1 by shearing. Thus, the Mohr-Coulomb model was presented to define failure when the shear stress in the material reaches the following criterion (see Figure 16.2):

$$\tau = c - \sigma \cdot \tan\phi \tag{16.1}$$

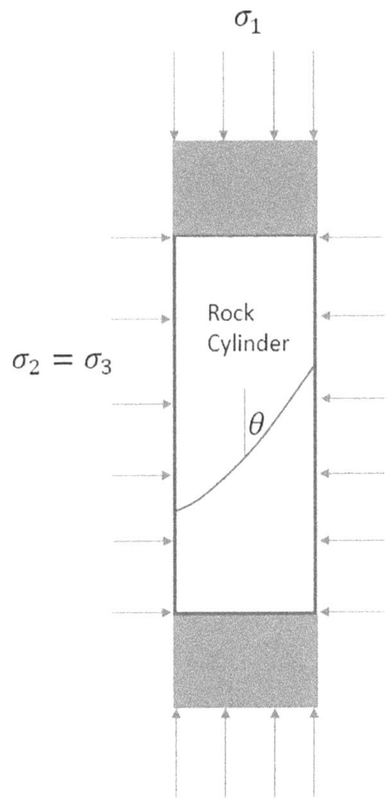

FIGURE 16.1 Soil under triaxial compression.

DOI: 10.1201/9781003436317-19

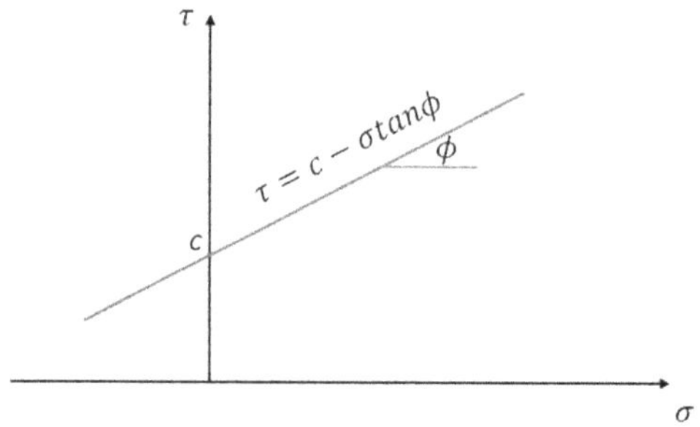

FIGURE 16.2 τ-σ of Mohr-Coulomb model.

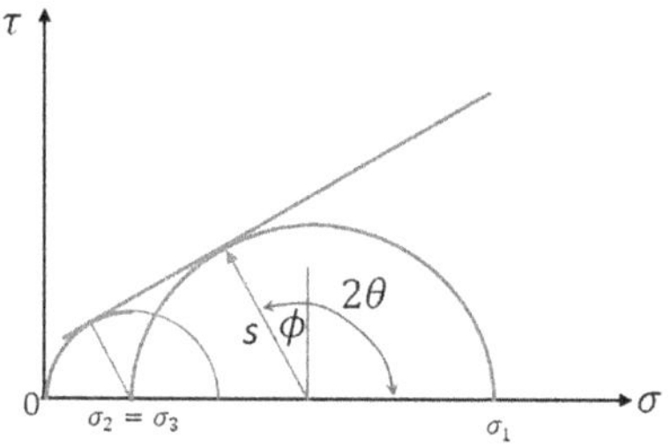

FIGURE 16.3 Mohr-Coulomb failure envelope.

in which τ – shear stress;

c – cohesion;
σ – normal stress; and
ϕ – inner friction angle.

The relationship between the angle θ and ϕ is illustrated in Figure 16.3:

$$\theta = 45^\circ + \frac{\phi}{2} \tag{16.2}$$

Also, Figure 16.3 shows that

$$s = \frac{\sigma_1 - \sigma_3}{2} \tag{16.3}$$

$$\tau = s \cdot \cos\phi \tag{16.4}$$

$$\sigma = \sigma_m + s \cdot \sin\phi \tag{16.5}$$

Substituting the above three equations into equation (16.1) results in the following:

$$s + \sigma_m \sin\phi - c \cdot \cos\phi = 0 \tag{16.6}$$

In the above equation, s is the maximum shear stress, and $\sigma_m = \frac{\sigma_1 + \sigma_3}{2}$ is the average of maximum stress and minimum stress. The yield surface of the Mohr-Coulomb model in π plane is a conical prism (see Figure 16.4).

16.2 MOHR-COULOMB MODEL IN ANSYS

Under triaxial compression, Figure 16.5a shows the true stress-strain curve of the soil, which can be approximated by a perfect plastic curve (see Figure 16.5b). Therefore, no hardening/softening law is required. At the yield surface,

$$f(\sigma_1, \sigma_2, \sigma_3) = 0 \tag{16.7}$$

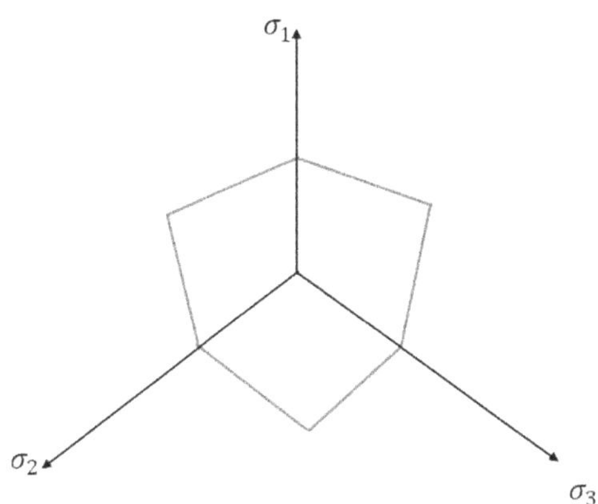

FIGURE 16.4 Yield surface of Mohr-Coulomb model in π plane.

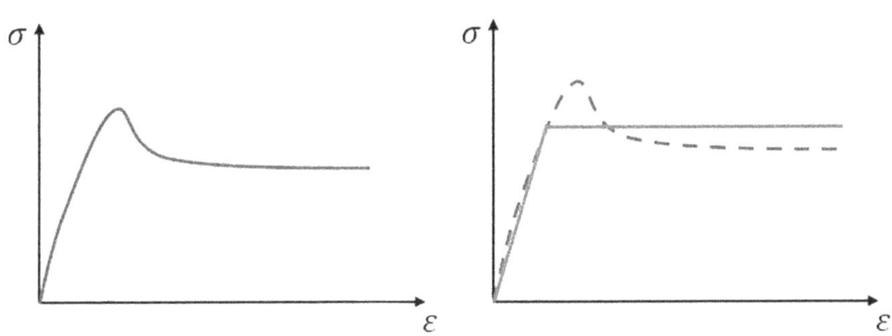

FIGURE 16.5 Real stress-strain curve of soil and its approximation. (a) Real soil. (b) Perfect plastic approximation.

For the flow rule of plastic strain,

$$\dot{\varepsilon}_i^p = \lambda \frac{\partial g}{\partial \sigma_i} \tag{16.8}$$

If $g = f$, it is classically associated plasticity; otherwise, it is general non-associated plasticity.

In ANSYS, the Mohr-Coulomb yield surface is defined as [1]

$$f_{MC}(\sigma) = \frac{\sigma_e}{\sqrt{3}}\left(\cos\vartheta - \frac{\sin\vartheta \sin\phi}{\sqrt{3}}\right) - \sigma\sin\phi - c\cos\phi \tag{16.9}$$

where

$$\sigma_e = \sqrt{3J_2} \tag{16.10}$$

$$\sin(3\vartheta) = -\frac{3\sqrt{3}}{2}\frac{J_3}{\sqrt{J_2^3}} \tag{16.11}$$

In tension, the yield surface of the Mohr-Coulomb model is specified as [1]

$$f_R(\sigma) = \frac{2}{3}\sigma_e \sin(\vartheta + 120°) - \sigma - T \tag{16.12}$$

where T is the tensile strength.

The Mohr-Coulomb flow potential is described by [1]

$$g_{MCt} = \frac{\sigma_e}{\sqrt{3}}\left(\cos\vartheta - \frac{\sin\vartheta \cdot \sin\psi}{\sqrt{3}}\right) - \sigma \cdot \sin\psi \tag{16.13}$$

where ψ is the dilatancy angle.

Therefore, the Mohr-Coulomb model needs to input the following material parameters in ANSYS [1]:

```
TB, MC, , , , BASE
TBDATA, 1,  ϕ, c, ψ , ϕ', c′
TB, MC, , , , RCUT
TBDATA, 1, T, T′
```

ϕ', c', and T' are the residual inner friction angle, residual cohesion, and residual tensile strength, respectively.

ANSYS also provides other optional parameters to define the softening of the Mohr-Coulomb model. Please check the ANSYS help documentation for details.

The Mohr-Coulomb model is widely used to represent aggregate materials such as soils, rocks, and concrete that start plastic deformation when the shear stress reaches a critical value. In the next sections, the Mohr-Coulomb model is applied to study the concrete slump test and the stability of a slope.

16.3 CONCRETE SLUMP TEST

Figure 16.4 shows that the yield surface of the Mohr-Coulomb model is not continuous, which makes convergence difficult during the Newton-Raphson solution.

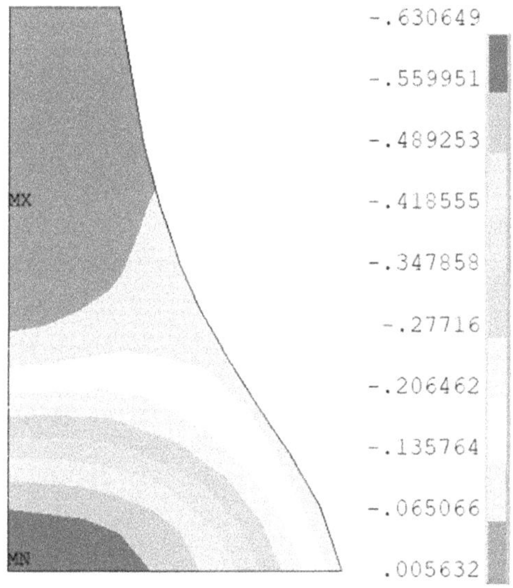

FIGURE 16.6 Plastic strains in the vertical direction.

To illustrate this point, the concrete slump test in Chapter 15.2 was repeated here using material data ($\phi = 20°$, $c = 0.1547e-3$) [2].

The solution reached convergence very slowly and required approximately 1,000 substeps to finish, compared with the 50 substeps of the Ducker-Prager model. The slow convergence of the Mohr-Coulomb model is due to its use of elastic material tangent. The plastic strains in the vertical direction that are plotted in Figure 16.6 differ from those in Figure 15.3 of Chapter 15.

16.4 STUDY OF SLOPE STABILITY

Landslides are often reported in the news. The slope stability associated with the landslides closely refers to the inclined soil's ability to withstand the slope movement. Some methods, such as kinematic analysis and limit equilibrium analysis, have been developed to study slope stability [3,4]. Due to advancements in technology in the past two decades, slope stability analysis has been investigated using the finite element method [5–7]. In this study, the slope stability of a slope was modeled by the Mohr-Coulomb model and studied using the finite element strength reduction method.

16.4.1 Finite Element Model

Half of a slope was developed in ANSYS190 (see Figure 16.7) because of its symmetry. The slope was meshed by PLANE182 (plane strain) with 1,050 elements and 1,116 nodes.

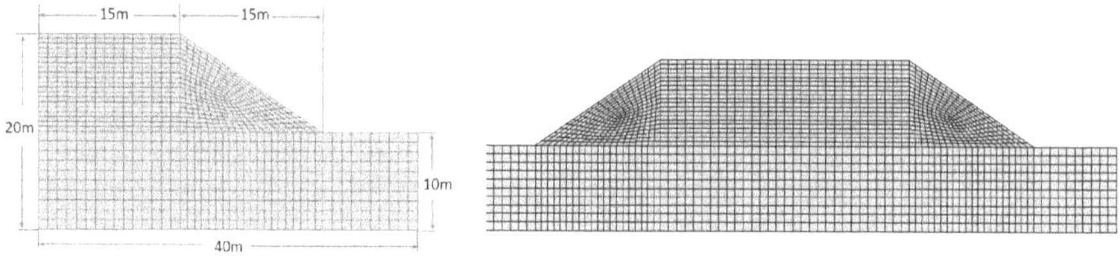

FIGURE 16.7 Finite element model of a slope. (a) Half a model. (b) Full model.

16.4.2 Material Properties

The slope was modeled with the Mohr-Coulomb material model, with different trial safety factors F_t listed in Table 16.1. The material parameters of the Mohr-Coulomb model with different F_t are calculated by

$$c_t = \frac{c}{F_t} \tag{16.14}$$

$$\phi_t = \arctan\left(\frac{\tan\phi}{F_t}\right) \tag{16.15}$$

The Mohr-Coulomb model was defined by the following:

```
TB,MC,1,,,BASE
TBDATA,1,17.2,10e3,0, 17.2,10e3  !c=15kPa, fictional angle 25°
TB,MC,1,,,RCUT
TBDATA,1,0.5e3,0.5e3
TB,MC,1,,,MSOL
TBDATA,1,2                        ! elastic tangent
```

16.4.3 Loadings and Boundary Conditions

The whole model was applied with self-weight by command ACEL, and the boundary of the slope was constrained (see Figure 16.8).

16.4.4 Results

Figure 16.9 plots the equivalent plastic strains with different trial safety factors F_t. The figure clearly shows that when $F_t = 1.4$, the plastic zone starts at the corner of the slope. When F_t increases to 1.6, the plastic zone gradually moves up. When $F_t = 1.63$, the plastic zone approaches the top of the slope. The whole plastic zone then looks like a curve. Thus, $F_t = 1.63$ should be the safety factor of the slope stability.

The deformation and von-Mises stresses of the slope when $F_t = 1.63$ are illustrated in Figures 16.10 and 16.11, respectively. Due to the loading of self-weight, the stresses of the slope increase with the increase in depth and reach their maximum at the bottom with a value of 0.216 MPa. The maximum displacement of 0.52 m occurs at the top.

TABLE 16.1
Material Parameters of Mohr-Coulomb Material Model

Ft	φ (◦)	*C* (kPa)
1.00	25	15
1.40	18.4	10.71
1.60	16.25	9.38
1.63	15.96	9.20

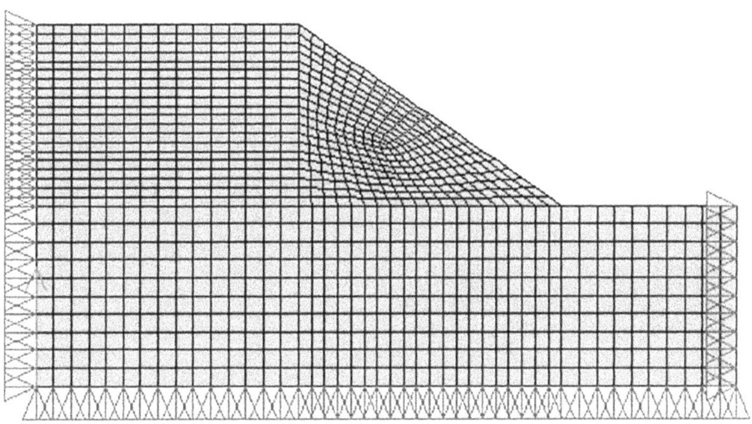

FIGURE 16.8 Loadings and boundary conditions.

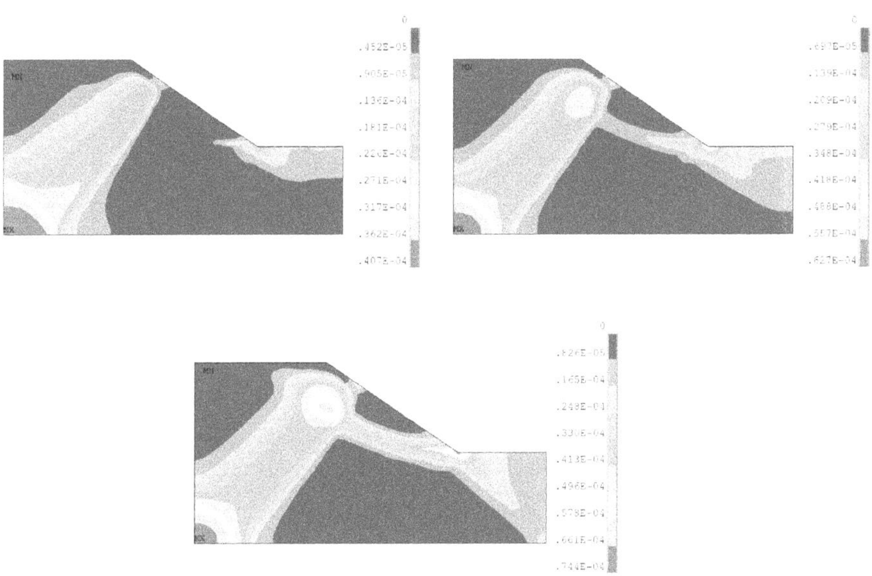

FIGURE 16.9 Equivalent plastic strains with different trial safety factors. (a) $F_t = 1.4$. (b) $F_t = 1.6$. (c) $F_t = 1.63$.

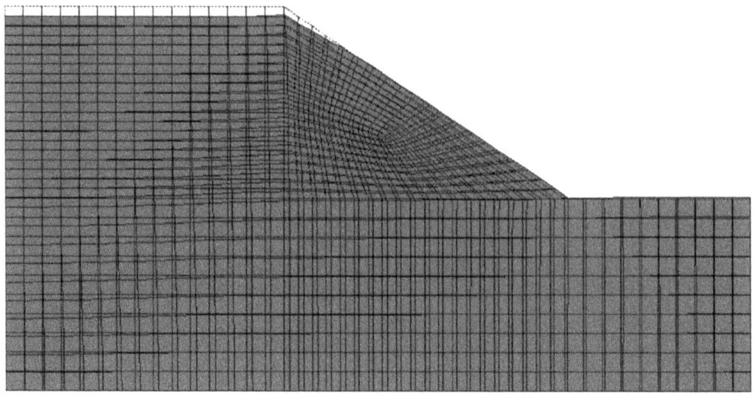

FIGURE 16.10 Deformation of the soil (Dmax = 0.52 m).

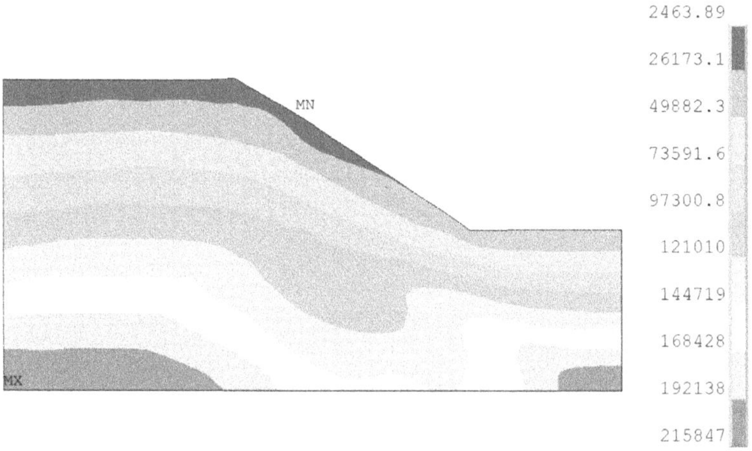

FIGURE 16.11 von-Mises stresses of the soil (F_t = 1.63) (Pa).

The typical convergence pattern of the Mohr-Coulomb material model is presented in Figure 16.12, which shows that it converges slowly and takes a lot of substeps to complete the entire computation.

16.4.5 Discussion

A 2D plane strain finite element model was created to study the stability of a slope. After comparing the plastic zones obtained by different trial safety factors, the safety factor of the slope was defined as the extent to which the obtained plastic zone extends from the bottom to the top of the slope.

In ANSYS, the Mohr-Coulomb and Jointed Rock models adopt the cutting-plane algorithm to solve the nonlinear equations because the Mohr-Coulomb and Jointed

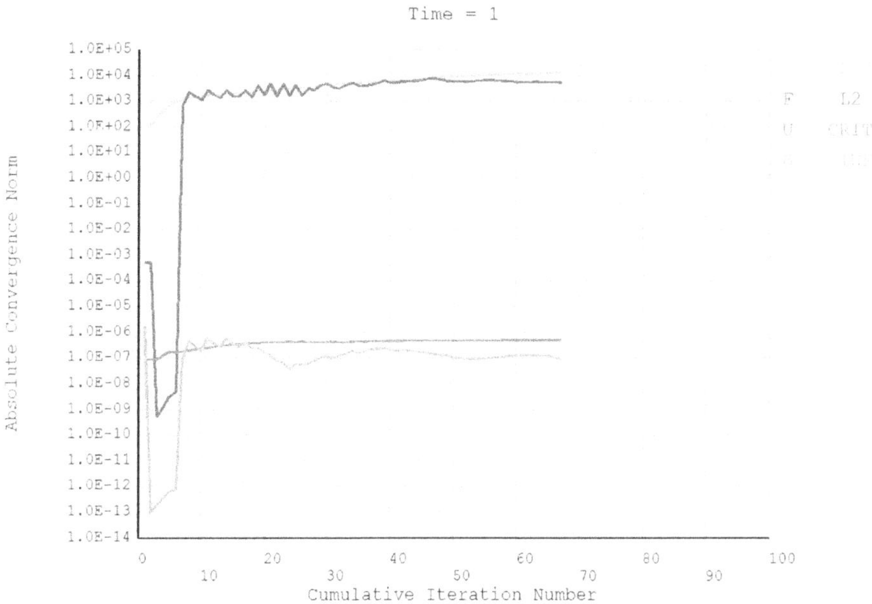

FIGURE 16.12 Convergence pattern of Mohr-Coulomb model.

Rock material models have a non-smooth yield surface. This cutting-plane algorithm has an elastic material tangent, which reduces the convergence speed of the Newton-Raphson iterations. That is why these material models always converge slowly and require a lot of substeps to complete the solution. In ANSYS, the cutting-plane algorithm can be selected in the material solution option using TB, MC, MSOL, in which options 1 and 2 refer to consistent tangent and elastic tangent, respectively.

16.4.6 Summary

The slope stability was studied with the Mohr-Coulomb model. The safety factor of the slope was defined by comparing the plastic zone obtained by different trial safety factors.

REFERENCES

[1] ANSYS190 Help documentation in the help page of product ANSYS190.
[2] ABAQUS version 6.6 documentation.
[3] Eberhardt, E, *Rock Slope Stability Analysis - Utilization of Advanced Numerical Techniques (PDF)*, Earth and Ocean Sciences, University of British Columbia, Vancouver, Canada, 2003.
[4] Bishop, A.W., The use of the slip circle in the stability analysis of slopes, *Geotechnique*, Vol. 5, 1955, pp. 7–17.
[5] Donald, I., and Chen, Z.Y. Slope stability analysis by the upper bound approach: fundamentals and methods, *Canada Geotechnical Journal*, Vol. 34, 1997, pp. 853–862.

[6] Griffiths, D.V., and Lane, P.A., Slope stability analysis by finite elements, *Geotechnique*, Vol. 49, 1999, pp. 387–403.
[7] Xu, X., and Dai, Z., Numerical implementation of a modified Mohr-Coulomb model and its application, *Journal of Modern Transportation*, Vol. 25, 2017, pp. 40–51.

17 Jointed Rock Model

Geologic and aggregate materials are inhomogeneous and contain weak planes in the materials, such as stratification planes. This kind of plane is named as a jointed rock, which is anisotropic and nonlinear, and has lower stiffness, higher permeability, and less strength than the intact rock. Various methods have been proposed to take the influence of discontinuities into account in the mechanical behavior of jointed rock [1–6]. Recently, ANSYS also released the Jointed Rock model for geomechanics simulation, which is introduced as follows.

17.1 JOINTED ROCK MODEL

The yield surface on a joint j is expressed in the form of the Mohr-Coulomb model as [7]

$$f_j = \tau_j - \sigma_j \tan\phi_j - c_j \tag{17.1}$$

where τ_j – shear stress on joint j,

σ_j – normal stress on joint j,
ϕ_j – frictional angle at joint j,
c_j – cohesion at joint j.

The flow potential for the joint is [7]

$$Q_j = \tau_j - \sigma_j \tan\psi_j \tag{17.2}$$

The tension-cutoff yield surface is defined as [7]

$$f_{Tj} = \sigma_j - T_j \tag{17.3}$$

where T_j is the tensile strength of joint j.

The residual yield surface and residual tensile strength yield surface follow equations (17.1) and (17.3), respectively, with different coefficients.

The orientation of the joint j separation plane is described by two angles (see Figure 17.1) [7]:

1. rotate the angle α around the Z axis clockwise and
2. rotate the angle β around the Y axis counterclockwise.

DOI: 10.1201/9781003436317-20

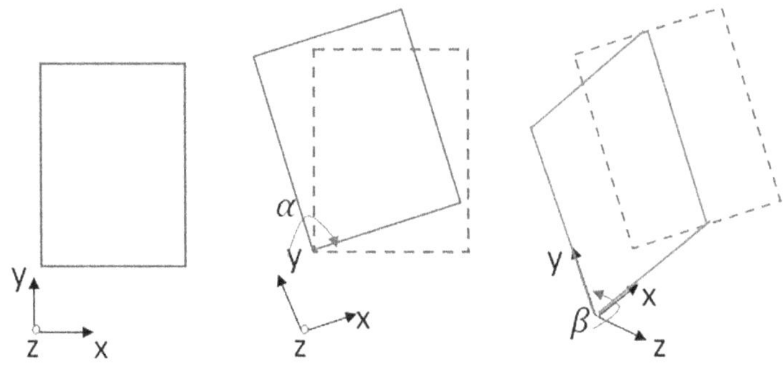

FIGURE 17.1 Determination of orientation of a joint [7]. (a) Initial position. (b) Rotate angle α clockwise around axis Z. (c) Rotate angle β counterclockwise around axis Y.

17.2 DEFINITION OF THE JOINTED ROCK MODEL IN ANSYS

The definition of the Jointed Rock model in ANSYS requires clarification of both the base material and the joints.

17.2.1 Defining the Base Material

The definition of the base material of the Jointed Rock model is the same as the one for the Mohr-Coulomb model [7]:

```
TB, JROCK,,,, BASE
TBDATA, 1, φ_j, c_j, ψ_j, φ'_j, c'_j
TB, JROCK,,,,RCUT
TBDATA,1, T, T'
TB, JROCK,,,,RSC
TBDATA, 1, R_SC
```

17.2.2 Defining the Joints

As many as four joints can be defined in ANSYS. Each joint demands yield stress and flow-potential parameters at the joint [7]:

```
TB, JROCK,,,,FPLANE
TBDATA, 1, φ_j, c_j, ψ_j, φ'_j, c'_j
TB, JROCK,,,,FTCUT
TBDATA,1, T, T'
```

Orientation of the joint is determined by the rotation angles α and β (see Figure 17.1) that are imported by the TB command:

```
TB, JROCK,,,,FORIE
TBDATA, 1, α, β
```

The above ANSYS commands can be repeated to define more joints.

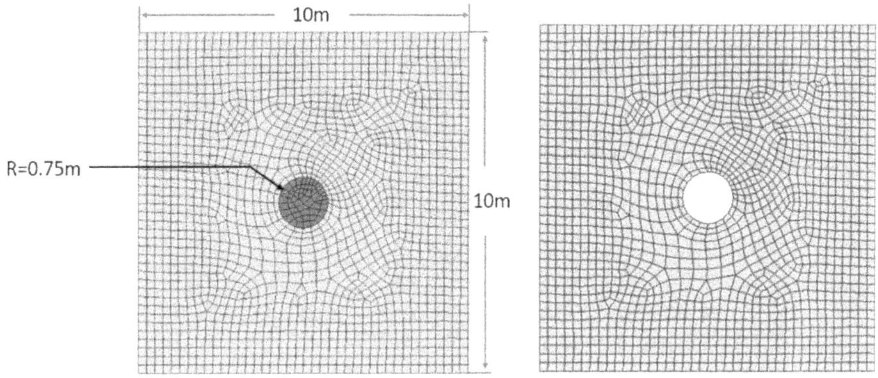

FIGURE 17.2 Finite element model of tunnel excavation. (a) Before excavation. (b) After excavation.

17.3 SIMULATION OF TUNNEL EXCAVATION

Tunnel construction requires excavation of the soil. Understanding the stress state of the soil during tunnel excavation is very important for the project design. In this study, the finite element method was applied to study the tunnel excavation using the Jointed Rock material model.

17.3.1 Finite Element Model

A tunnel with a diameter of $D = 1.5$ m was excavated from a 10 m width square [8]. Thus, the whole model was composed of the tunnel (purple in Figure 17.2) and the square without the tunnel (blue in Figure 17.2) and meshed by PLANE182 (plane strain) with 1,669 elements and 1,749 nodes.

17.3.2 Material Properties

The area around the tunnel was modeled with the Jointed Rock material model. The base material, regarded as the Mohr-Coulomb material, was defined by

```
TB,JROCK,1,,,BASE
TBDATA,1,21.2,1e6,10,21.2,1e6
TB,JROCK,1,,,RCUT
TBDATA,1,0.1e6,0.1e6

TB,JROCK,1,,,RSC
TBDATA,1,0
```

The Jointed Rock, including its material model and orientation of the joint failure plane, was specified by

```
TB,JROCK,1,,,FPLANE
TBDATA,1,15.8,0.1e6,0,15.8,0.1e6
```

```
TB,JROCK,1,,,FTCUT
TBDATA,1,0.05e6,0.05e6

TB,JROCK,1,,,FORIE
TBDATA,1,,-60
```

17.3.3 Loadings and Boundary Conditions

The boundary of the whole model was fixed (see Figure 17.3), and the initial stress of 1e6 Pa was applied to all the elements in both *X* and *Y* directions by

```
INISTATE,SET,DTYP,STRE
INISTATE,DEFI,ALL,,,,-1E6,-1E6,,0,0,0
```

17.3.4 Solution

The solution was completed in two steps. The first step solved the full square with the initial stress. In the second step, the elements within the central tunnel part were deactivated using the EKILL command to simulate excavation.

17.3.5 Results

Figure 17.4 illustrates the final deformation of the model. The displacements mainly occur around the tunnel, with a maximum displacement of 7.35 mm. The von-Mises

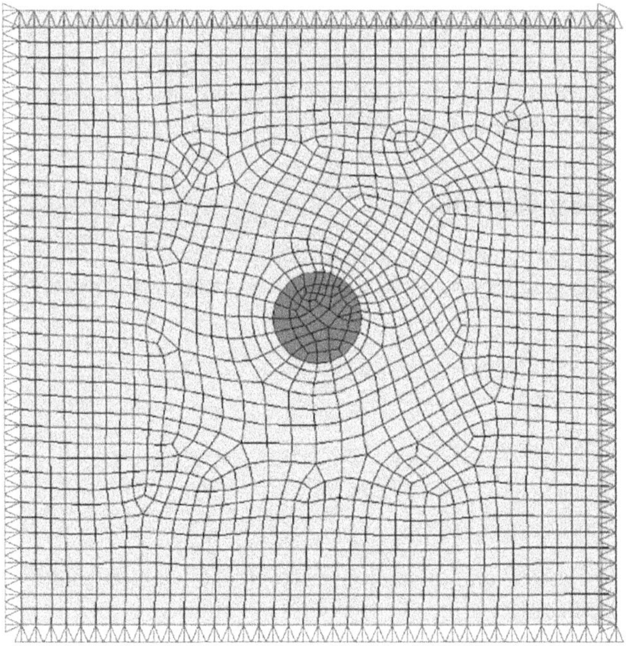

FIGURE 17.3 Boundary conditions.

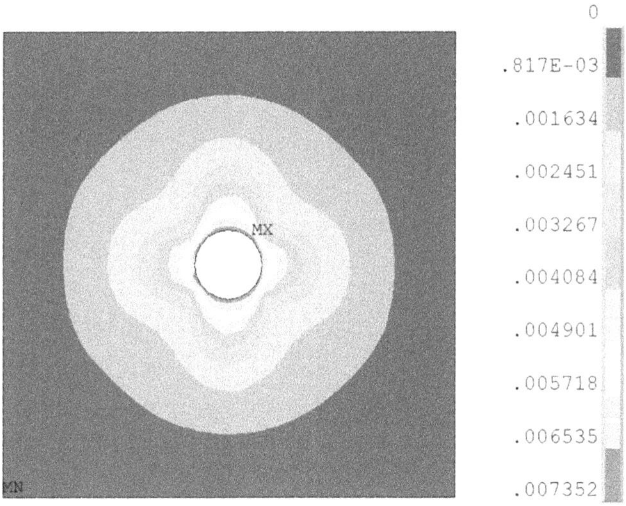

FIGURE 17.4 Final deformation (m).

FIGURE 17.5 Stresses of the soil at the end of the first step (stress = 847,950 Pa).

stresses of the model are plotted in Figures 17.5 and 17.6. In the first step, the stress distribution is uniform at 0.85 MPa. The stress concentration occurs around the edge of the tunnel in the second step because of the excavation. The final plastic strain contour is presented in Figure 17.7, which clearly shows that the main plastic strains occur in

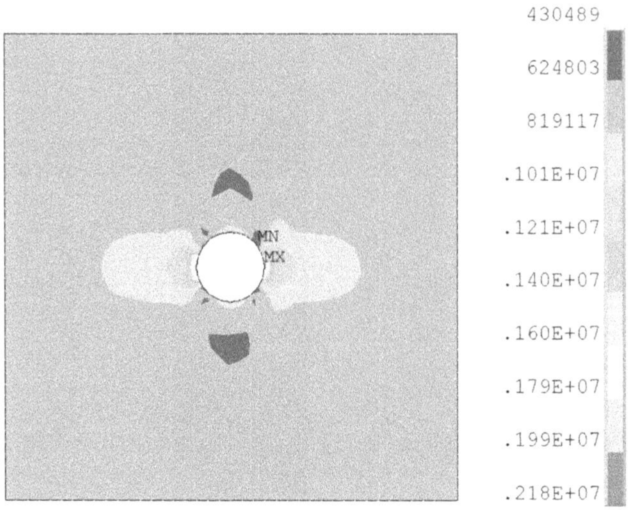

FIGURE 17.6 Final von-Mises stresses of the soil (Pa).

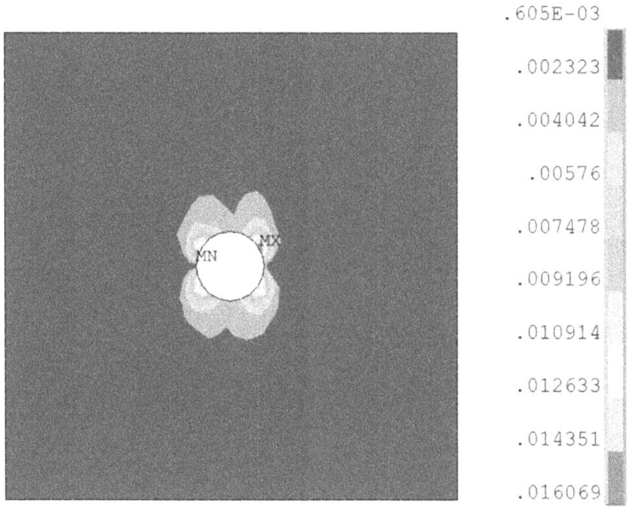

FIGURE 17.7 Final von-Mises plastic strains of the soil.

four different directions. For comparison, the final von-Mises stresses of the model with the Mohr-Coulomb material model are given in Figure 17.8, showing that the stresses distribute uniformly around the edge of the tunnel and that no plastic strain occurs.

17.3.6 Discussion

A 2D plane strain finite element model was developed to simulate the tunnel excavation using the Jointed Rock material model. After excavation, the stress concentration occurs at the edge of the tunnel, and the major plastic stains are in four different

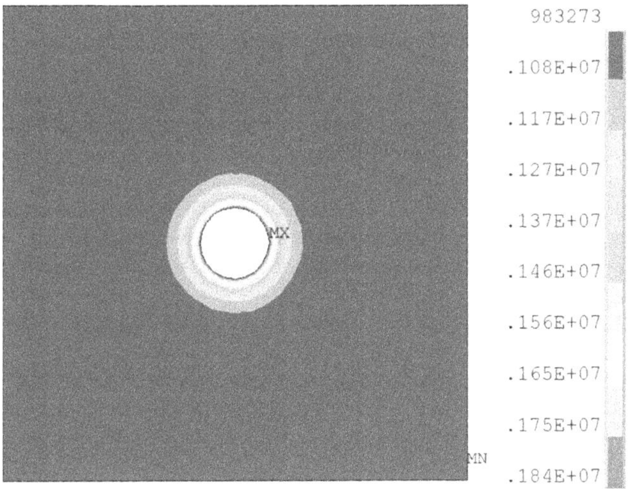

FIGURE 17.8 Final von-Mises stresses of the soil with Mohr-Coulomb model only (Pa).

directions, which match the behavior of the Jointed Rock material model. The same model without specifying the failure plane shows a uniform stress distribution around the edge of the tunnel and no plastic stain. The different results state clearly that the stress concentration and the major plastic stains are due to the failure plane of the Jointed Rock material model.

The EKILL command was applied to deactivate the elements to simulate excavation. Element birth and death is very useful in modeling. For example, contact element birth and death is used in simulation of the contact between the stent and the vessel in angioplasty [9]. The same method is adopted in Chapter 21 to simulate the contact between the rod and the flange. In addition, initial stress was loaded to simulate the stress state at the beginning, which is another advanced finite element technology of ANSYS.

The whole computation was completed in two steps. The first step was elastic. Therefore, it was solved in one substep. Excavation in the second step increased the stress state of the soil beyond the elastic stage and triggered material nonlinearity. Thus, it took 11 substeps to complete. Therefore, different substeps were selected for the two steps.

17.3.7 Summary

The tunnel excavation was simulated using the Jointed Rock material model. The computational results match the behavior of the material model.

REFERENCES

[1] Ramamurthy, T., A geo-engineering classification for rocks and rock masses, *International Journal of Rock Mechanics and Mining Sciences*, Vol. 41, 2004, pp. 89–101.

[2] Jaeger, J., Shear failure of anistropic rocks, *Geological Magazine*, Vol. 97, 1960, pp. 65–72.

[3] Chen, W., Konietzky, H., and Abbas, S.M., Numerical simulation of time-independent and-dependent fracturing in sandstone, *Engineering Geology*, Vol. 193, 2015, pp. 118–131.
[4] Chakraborti, S., Konietzky, H., and Walter, K., A comparative study of different approaches for factor of safety calculations by shear strength reduction technique for non-linear Hoek-Brown failure criterion, *Geotechnical and Geological Engineering*, Vol. 30, 2012, pp. 925–934.
[5] Brzovic, A., and Villaescusa, E., Rock mass characterization and assessment of block-forming geological discontinuities during caving of primary copper ore at the El Teniente mine, Chile, *International Journal of Rock Mechanics and Mining Sciences*, Vol. 44, 2007, pp. 565–583.
[6] Bray, J., A study of jointed and fractured rock, *Rock Mechanics and Engineering Geology*, Vol. 5, 1967, pp. 117–136.
[7] ANSYS190 Help Documentation in the help page of product ANSYS190.
[8] Chang, L., and Konietzky, H., Application of the Mohr-Coulomb yield criterion for rocks with multiple joint sets using fast Lagrangian analysis of continua 2D (FLAC2D) software, *Energies*, Vol. 11, 2018, p. 614.
[9] Yang, Z., *Finite Element Analysis for Biomedical Engineering Applications*, CRC Press, Boca Raton, 2019.

18 Consolidation of Soils

Chapter 18 discusses the consolidation of soils. Section 18.1 introduces the consolidation of soils, Section 18.2 presents the simulation of porous media in ANSYS, and Sections 18.3 and 18.4 focus on the simulation of the consolidation.

18.1 CONSOLIDATION OF SOILS

Soil primarily consists of solids and water. When the soil is under compression, the volumetric stiffness of water is much higher than that of the solid; therefore, the water absorbs the pressure and flows away from the region of high pressure. Thus, the solid gradually takes over the loading and shrinks in volume. That is the process of soil consolidation.

When a building is erected over a layer of soil with low permeability and low stiffness, a large settlement of the building may occur over many years due to the consolidation. Also, soil consolidation significantly influences construction projects such as tunnel excavations and the construction of embankments. Therefore, it is necessary to conduct a numerical simulation of soil consolidation. ANSYS provides strong tools in this field, which are introduced in the following section.

18.2 MODELING POROUS MEDIA IN ANSYS

Porous media are regarded as two-phase materials governed by an extended Biot's consolidation theory [1]. The fluid flow is described by Darcy's law as

$$\mathbf{q} = -k\nabla p \tag{18.1}$$

where p – pore pressure,

k – permeability, and
∇ – gradient operator.

The solid phase follows the equation

$$\nabla \cdot \left(\boldsymbol{\sigma}'' - \alpha \cdot p\mathbf{I}\right) + f = 0 \tag{18.2}$$

where f – body force,

α – Biot coefficient,
$\boldsymbol{\sigma}''$ – Biot effective stress, and
$\mathbf{I}$ – second-order identity tensor.

DOI: 10.1201/9781003436317-21

Based on equations (18.1) and (18.2), ANSYS developed Coupled Pore-Pressure-Thermal (CPT) elements such as CPT212 (2D 4-node), CPT213 (2D 8-node), CPT215 (3D 8-node), CPT216 (3D 20-node), and CPT217 (3D 10-node) to simulate the porous media. The material properties of porous media are defined by [2]

```
TB, PM,,,, PERM
TBDATA, 1, 1e-8                 ! define solid permeability k
TB, PM,,,, BIOT               ! define Biot coefficient
TBDATA, 1, 0.5
```

18.3 SIMULATION OF TERZAGHI'S PROBLEM

Terzaghi's one-dimensional consolidation problem is a standard problem in the field of geomechanics [3]. Its simulation may provide insight into the process of soil consolidation.

Terzaghi's problem was modeled in ANSYS190 using the porous media material model with CPT212 (plane strain option). A 12-meter-long soil was modeled with the porous material defined as Young's modulus 6,000 Pa and permeability 4e-6 m (see Figure 18.1). The bottom of the soil was fixed, and the top was loaded with 1 Pa. In addition, the pore pressure of the top, which was defined as zero, allowed water to escape. Figure 18.2 shows the change of the pore pressure of the bottom over time; this indicates that while the water expels from the top under pressure, the pore pressure gradually decreases and eventually approaches constant.

18.4 SIMULATION OF CONSOLIDATION OF THREE-WELL ZONE

Primary consolidation is the settlement process in which a vertical stress changes due to an applied load that causes water to be squeezed out of the soil. The process is complete when the soil loses all its water. In this study, the consolidation was simulated in ANSYS190 using CPT elements.

18.4.1 Finite Element Model

The whole area consists of three rock zones and two fault zones (see Figure 18.3). Three wells with a radius of 0.2 m are distributed uniformly within the middle rock zone. The top and bottom rock zones, as well as the fault zones, were modeled with the elastic material and meshed by PLANE182 in the plane strain state. The central soil zone was regarded as porous media and meshed with CPT212. The entire model consists of 3,995 elements and 4,151 nodes (see Figure 18.4).

18.4.2 Material Properties

The top and bottom rock zones were assumed to be linear elastic according to Young's modulus of 65 GPa and Poisson's ratio of 0.25. The middle rock zone was modeled as porous media, with the material properties defined as follows:

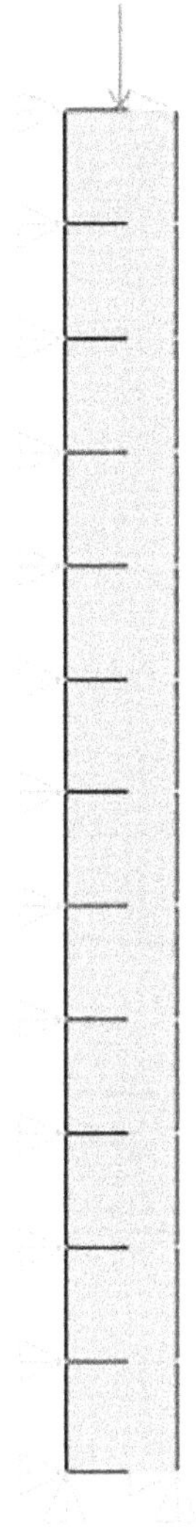

FIGURE 18.1 One-dimensional Terzaghi's model.

```
FPX=1.0D-10
ONE=1.0

TB,PM,2,,,PERM
TBDATA,1, FPX, FPX, FPX
TB, PM,2,,,BIOT                    ! DEFINE BIOT COEFFICIENT
TBDATA, 1, ONE
```

The fault zone with Young's modulus of 1 GPa was much softer than the rock zones.

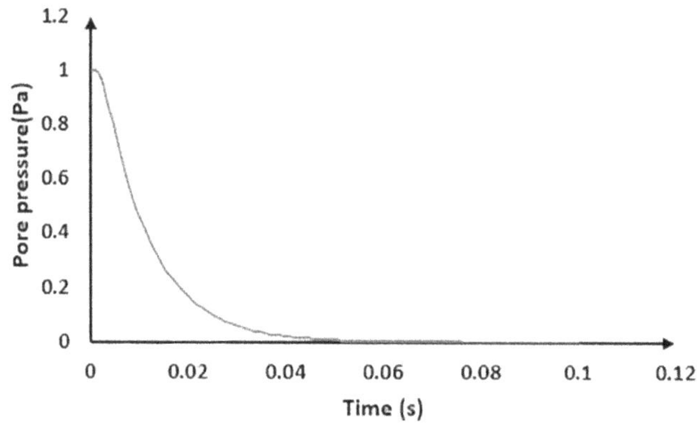

FIGURE 18.2 Variation of pore pressure at the end with time.

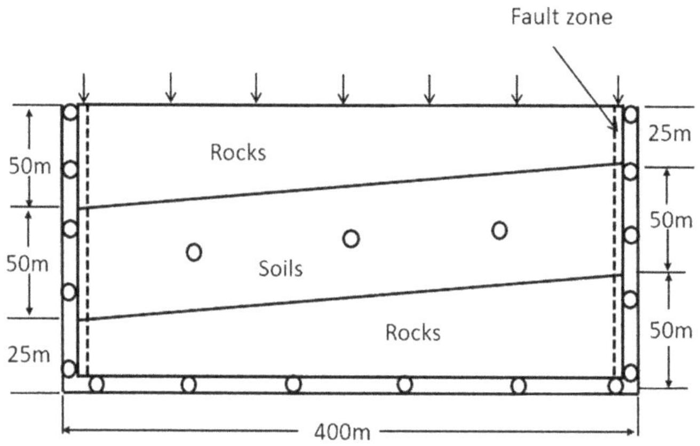

FIGURE 18.3 Consolidation of three-well zone.

18.4.3 Boundary Conditions and Loadings

The whole area was constrained in the normal directions of the boundary (see Figure 18.5).

The total computation consisted of two steps. In the first step, pressures of 100 and 50 MPa were loaded on the top surface and the well wall, respectively. Because the well walls had no flow, no pore pressure was defined in the first step. However, the pore pressure of the well wall surfaces changed to zero in the second step.

18.4.4 Solutions

The pressure loading in the first step was ramped (KBC, 0), and the loading in the second step was stepped (KBC, 1). Time in the first and second steps was set at 1 second and 3 hours, respectively.

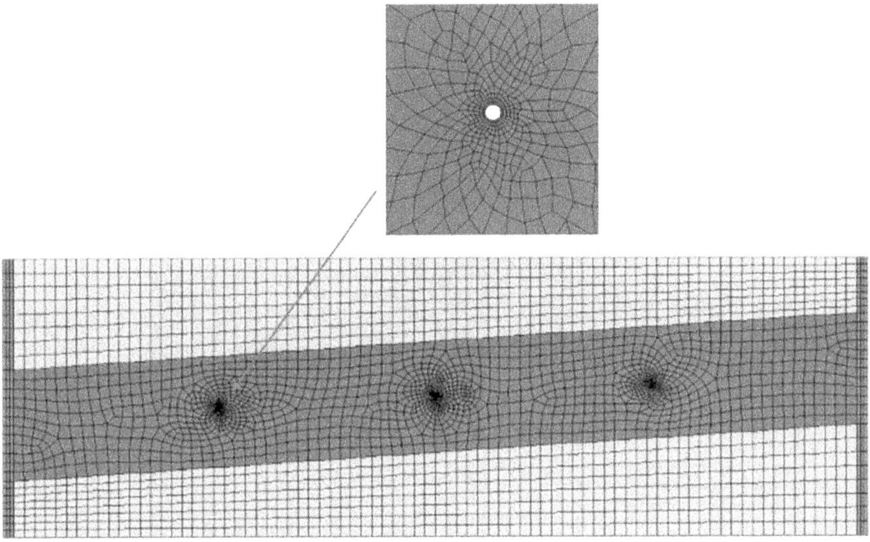

FIGURE 18.4 Finite element model of three-well zone.

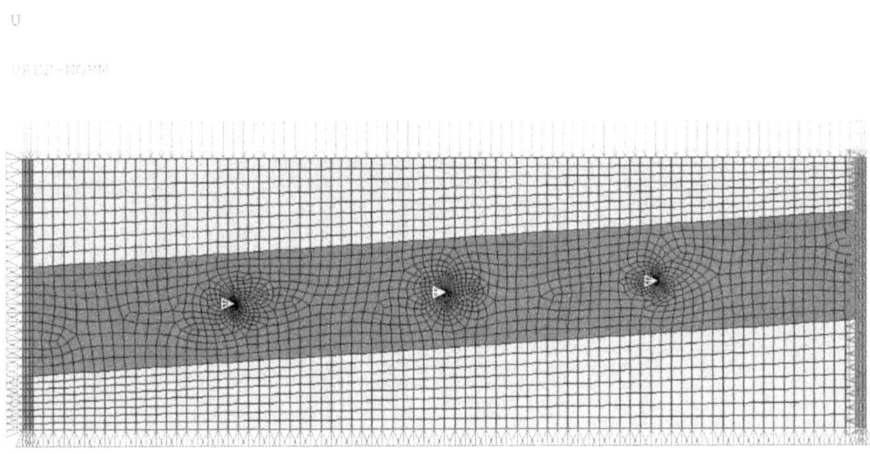

FIGURE 18.5 Boundary conditions and loadings.

18.4.5 Results

Figure 18.6 shows the final deformation of the whole model. The displacement decreases with the depth, and the maximum deformation of 0.88 m occurs at the top of the fault zone because the fault zone is much softer than the rock zones. Figure 18.7 plots the final von-Mises stresses. The stress is approximately uniformly distributed except for the top interface between the rock and fault zone. This is because at the final time, the water in the rock had almost completely been expelled into the well,

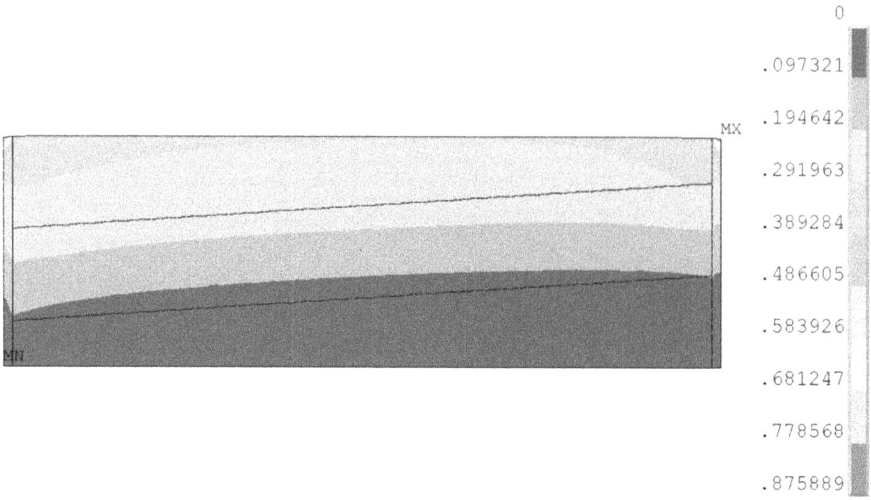

FIGURE 18.6 Final deformation of the model (m).

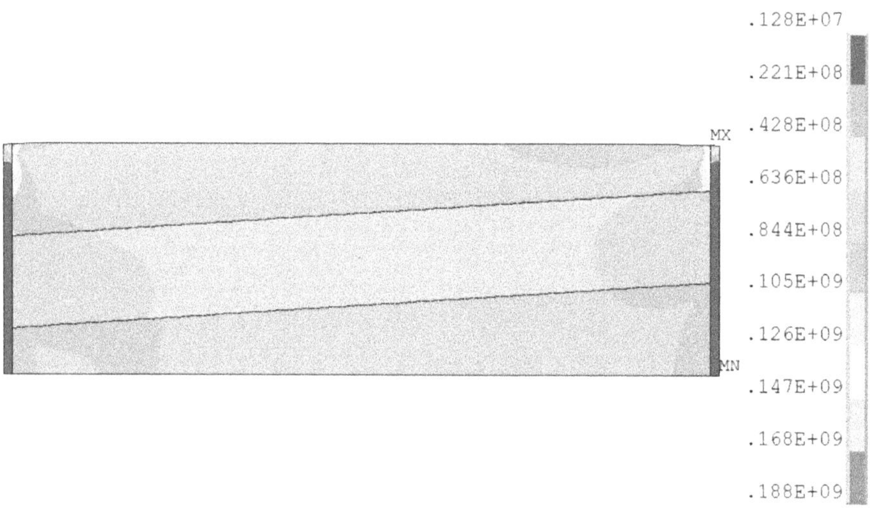

FIGURE 18.7 Final von-Mises stresses of the model (Pa).

and the rock zone had become purely solid. Under pressure loading, the stresses in the rock should be close within the rock. The pore pressures of the model at the end of the first and second steps are presented in Figures 18.8 and 18.9, respectively; they clearly indicate that after 3 hours, the pore pressures dropped from about 98 to 1.0 MPa. That is confirmed by the time history of two nodes within the rock zone (see Figure 18.10). Node A is closer to the well than node B. That is why the pore pressure of node A initially drops faster than that of node B. After 6,000 seconds, both are nearly the same.

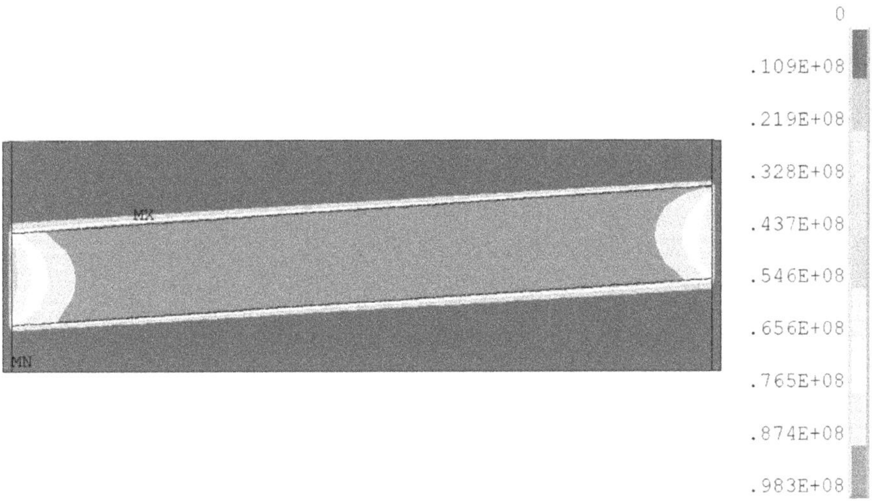

FIGURE 18.8 Pore pressures of the model at the end of the first step (Pa).

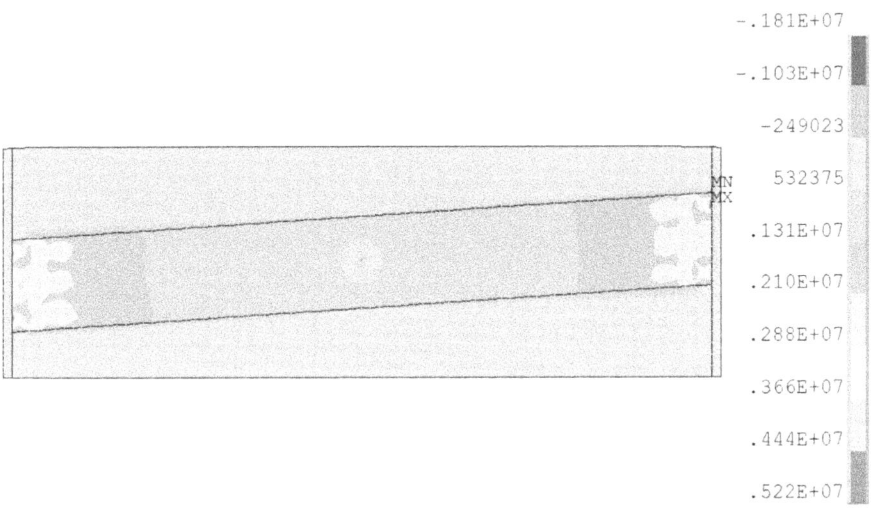

FIGURE 18.9 Final pore pressures of the model (Pa).

18.4.6 Discussion

A 2D plane strain finite element model was created to simulate the consolidation of the rocks using the pore media material model. After 3 hours, the water was expelled into the well, and the pore pressure within the rock dropped of 99%. This model confirms that CPT elements can be used to simulate soil consolidation.

Compared to SOLID elements such as SOLID185, the CPT elements have one extra DOF, pore pressure. Zero pore pressure was applied to define the water flow at

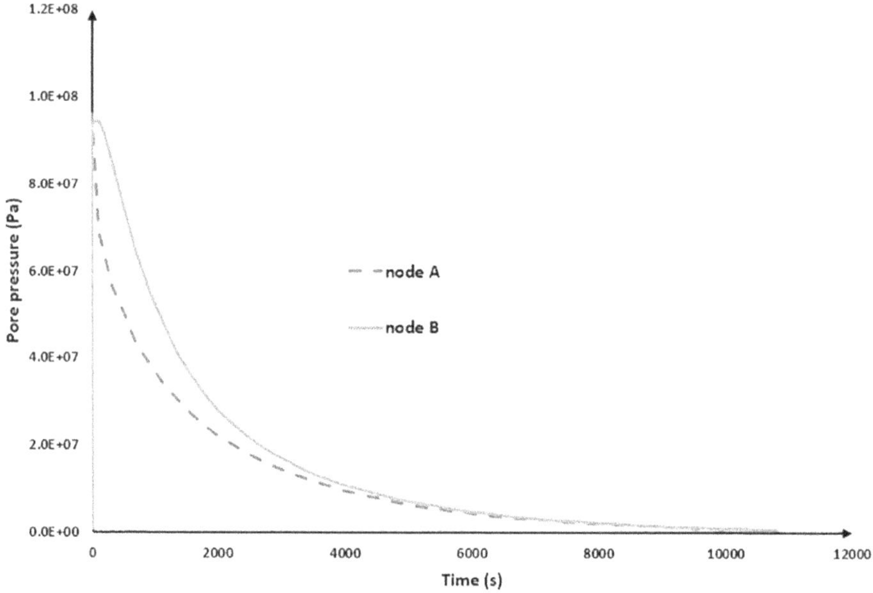

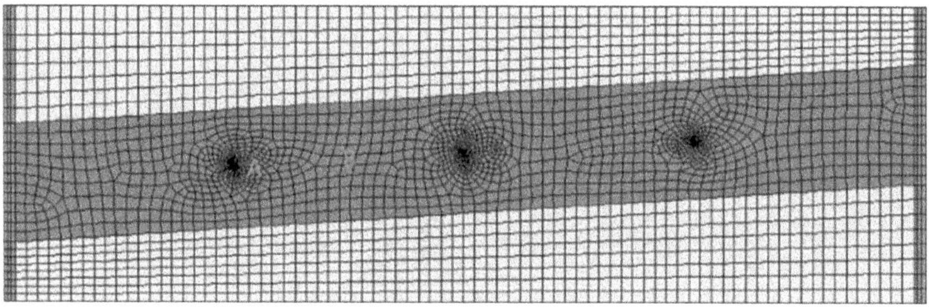

FIGURE 18.10 Time history of pore pressures of nodes A and B. (a) Time history of pore pressures. (b) Locations of nodes A and B.

the boundary. If no pore pressure is defined in the boundary, then the boundary has no water flow. Negative pore pressure refers to the area with a suction effect.

18.4.7 Summary

The consolidation of the rocks was studied using CPT elements in ANSYS. After a long time, the pore pressure dropped to near zero because the water had been expelled into the well.

REFERENCES

[1] Biot, M.A., General theory of three-dimensional consolidation, *Journal of Applied Physics*, Vol.12, 1941, pp. 155–164.
[2] ANSYS190 Help Documentation in the help page of product ANSYS190.
[3] Terzaghi, K., *Theoretical Soil Mechanics*. John Wiley and Sons, New York, 1943.

Part IV

Modern Materials

In addition to the traditional materials discussed in the previous three parts, some modern materials, such as composites, shape memory alloys, and functionally graded materials, have been developed and widely used in military aircraft, space, the automotive industry, and biomedical fields. The mechanical design of these materials demands an understanding of the stress and strain states of these materials, which can be achieved using the finite element method. Therefore, learning how to model these modern materials in finite element analysis becomes vital. That is the topic of Part IV.

Chapter 19 talks about composites, including how to model them in ANSYS, and provides two examples related to composite damage evolution and crack growth. Functionally graded materials are simulated in Chapter 20 using TBField technology in ANSYS. Chapter 21 focuses on shape memory alloys and covers their application for orthodontic wire and a vacuum-tight shape memory flange. Piezoelectric materials, as well as the simulation of a thin-film piezoelectric microaccelerometer, are presented in Chapter 22. The final chapter explains how to determine the Young's modulus of Fe nanoparticles from the experimental data in ANSYS.

DOI: 10.1201/9781003436317-22

19 Composite Materials

Composite materials have a long history that can be traced back to the Israelites using straw to strengthen mud bricks. Modern composite materials were originally developed in the 1950s to address aerospace's demand for materials with light weight, high strength, long fatigue life, and good thermal conductivity. Now, composite materials have been widely applied in military aircraft, civil aircraft, space, and the automotive industry.

Chapter 19 briefly introduces composite materials, describes modeling composite materials in ANSYS, and offers two applications of composite materials.

19.1 INTRODUCTION OF COMPOSITE MATERIALS

Composite material signifies that two or more materials are combined to form a third material. Unlike the combination on a microscopic scale, like the alloying of metals that is homogeneous macroscopically, the combination of materials in composite materials, which can be identified by the naked eye, is non-homogeneous on a macroscopic scale.

The composite materials are classified as

- Laminated composite materials consisting of layers of various materials (see Figure 19.1);
- Fibrous composite materials consisting of fibers in a matrix (see Figure 19.2);
- Particulate composite materials composed of particles in a matrix; and
- A combination of the above three types.

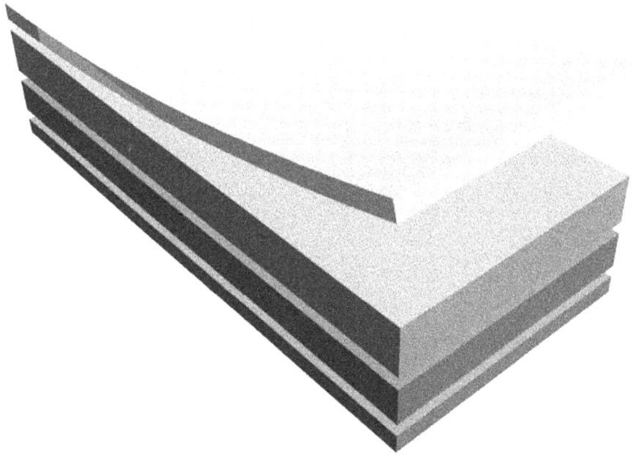

FIGURE 19.1 Illustration of the laminated composite. (snehit©123rf.com.)

DOI: 10.1201/9781003436317-23

FIGURE 19.2 Fibrous composite material. (coddie©123rf.com.)

Well-designed composite materials usually exhibit better properties than any of their components or constituents. These properties improved by forming a composite material include, but are not limited to, the following [1]:

- Strength,
- Stiffness,
- Fatigue life,
- Corrosion resistance,
- Wear resistance,
- Thermal conductivity,
- Acoustical insulation, and
- Weight.

The composite materials can be designed to improve some of the above properties, but not all of them at the same time. Even some of the properties conflict with each other, such as thermal conductivity versus thermal insulation. Thus, the goal of designing the composite materials is to enhance the specific properties required by the projects.

The composite materials also have some disadvantages [1]:

- Delamination – Composites always delaminate between layers;
- Damage inspection – Cracks and delamination in composites are usually internal and require special inspection techniques for detection;
- Complex fabrication – The fabrication of composites is always labor intensive and complex;

- High cost – Because the composite materials are new, they have a very high cost; and
- Composite to metal joining – The imbalance at joinery is due to the different thermal properties of the metal and composite and may lead to failure.

As composites play such an important role in modern industry, many features are available in ANSYS for modeling composites. They are introduced in the next section.

19.2 MODELING COMPOSITE IN ANSYS

19.2.1 Modeling Composite by Command SECTYPE

Many layers of fibrous composite materials are assembled to form a composite (see Figure 19.1). To define one layer, the following properties need to be specified:

a. Layer thickness (TK),
b. Material properties (MAT), and
c. Layer orientation angle commands (THETA).

The above parameters can be imported into ANSYS [2] as

```
SECTYPE,1,SHELL
SECDATA,TK1,MAT1,THETA1 ! THICKNESS, MATERIAL ID, ANGLE OF 1ST
LAYER
SECDATA,TK2,MAT2,THETA2 ! THICKNESS, MATERIAL ID, ANGLE OF 2ND
LAYER

...

SECDATA,TKN,MATN,THETAN ! THICKNESS, MATERIAL ID, ANGLE OF NTH
LAYER
```

This defines composition, while the following explains that composite damage can be characterized with damage-initiation criteria and damage-evolution laws. The damage-initiation criteria are defined in ANSYS as

```
TB,DMGI,1,1,,FCRT  ! FC for damage initiation
TBDATA, 1, C1 ! Failure criteria type for the tensile fiber
failure mode
TBDATA, 2, C2 ! Failure criteria type for the compressive
fiber failure mode
TBDATA, 3, C3 ! Failure criteria type for the tensile matrix
failure mode
TBDATA, 4, C4 ! Failure criteria type for compressive matrix
failure mode
```

The damage-evolution law is specified in ANSYS by [2]

```
TB,DMGI,1,1,,MPDG  ! FC for damage evolution
TBDATA, 1, C1 ! Tensile fiber stiffness reduction
TBDATA, 2, C2 ! Compressive fiber stiffness reduction
TBDATA, 3, C3 ! Tensile matrix stiffness reduction
TBDATA, 4, C4 ! Compressive matrix stiffness reduction
```

19.2.2 Modeling Composite by Anisotropic Model

Unlike the most common engineering materials, composite materials are anisotropic. For a point of the body, anisotropic materials have material properties that differ with the direction. One special kind of anisotropic material is the orthotropic material with different material properties in three mutually perpendicular direction planes, which is defined in ANSYS using MP commands [2]:

```
MP, EX, ,E1               ! elastic modulus, element x direction
MP, EY, ,E2               ! elastic modulus, element y direction
MP, EZ, ,E3               ! elastic modulus, element z direction
MP, PRXY/NUXY, ,   υ12 ! major (minor) Poisson's ratio, x-y plane
MP, PRYZ/NUYZ , ,  υ23 ! major (minor) Poisson's ratio, y-z plane
MP, PRXZ/NUXZ , ,  υ13 ! major (minor) Poisson's ratio, x-z plane
MP, GXY, , G12            ! shear modulus, x-y plane
MP, GYZ, , G23            ! shear modulus, y-z plane
MP, GXZ, , G13            ! shear modulus, x-z plane
```

As a simple way to specify the composite materials, this definition is widely used in the simulation of crack growth in composites.

19.3 SIMULATION OF COMPOSITE STRUCTURE IN FAILURE TEST

The composite structure failure test is critical in determining the applied flight-load levels on a composite structure. It is performed until structural failure occurs [3]. The aim of this study is to simulate the composite failure in ANSYS, which is helpful for understanding the failure mechanism of the composite structure.

19.3.1 Finite Element Model

The composite structure is composed of an adapter, load head, and conic part. The adapter connects the load head and conic part (see Figure 19.3). The conic part is a composite with a sandwich construction, in which the inner and outer composite faces are a $[(90/0)_4]$ layup. The finite element model of the composite structure was created in ANSYS190 (see Figure 19.4), in which the adapter and load head as well as the solid section of the conic part were meshed with SOLID185 and the composite section of the conic part was meshed with SHELL181. The whole model consists of 30,983 nodes and 130,395 elements.

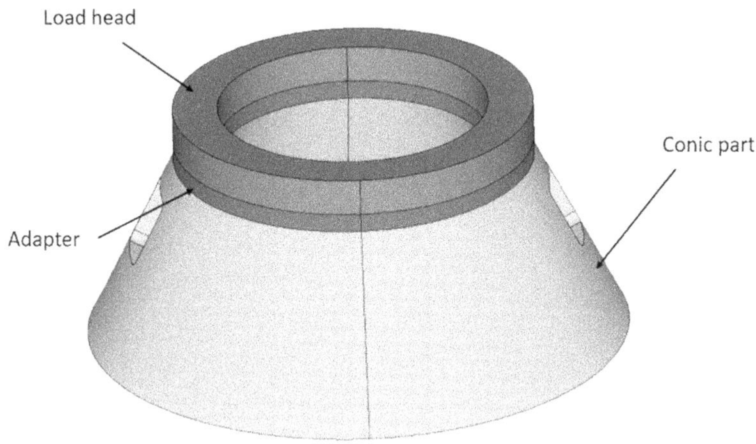

FIGURE 19.3 Structure of composite structure failure test.

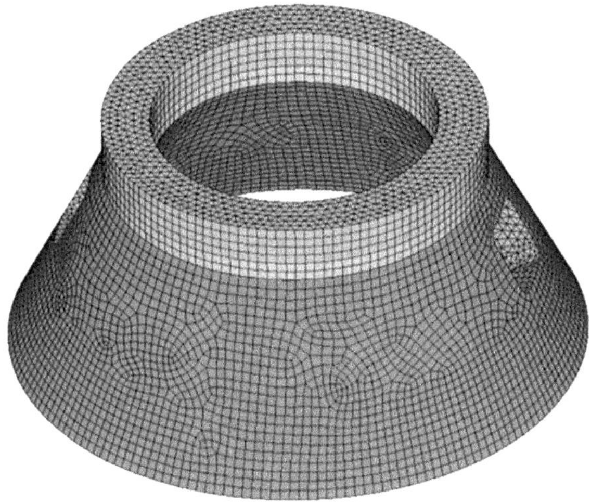

FIGURE 19.4 Finite element model of the test.

19.3.2 Material Properties

The adapter and load head were assumed to be linear elastic with a Young's modulus of 180 GPa and a Poisson's ratio of 0.3. Each composite ply of the conic part is 1.5 mm thick and composed of carbon/epoxy AS4-3501-6. The core of the sandwich composite is Rohacell 110 WF with a thickness of 25 mm. Rohacell 110 WF was assumed to be linear elastic with a Young's modulus of 180 MPa and a Poisson's ratio of 0.286. AS4-3501-6 was modeled as orthotropic [3]:

```
!TYPICAL MATERIAL PROPERTIES FOR E-GLASS/EPOXY (MPA)
E_X=18.4*6894
E_Y=1.62*6894
E_Z=E_Y
PR_XY=0.1279
PR_YZ=0.1331
PR_XZ=PR_XY
G_XY=0.951*6894
G_YZ=0.528*6894
G_XZ=G_XY
MP,EX,1,E_X
MP,EY,1,E_Y
MP,EZ,1,E_Z
MP,NUXY,1,PR_XY
MP,NUYZ,1,PR_YZ
MP,NUXZ,1,PR_XZ
MP,GXY,1,G_XY
MP,GYZ,1,G_YZ
MP,GXZ,1,G_XZ
```

The composite faces are a [(90/0)4], in which 0° and 90° are in the axial direction and the hoop direction, respectively. Therefore, a local coordinate system 11 is defined with *X* in the vertical direction as the element coordinate system of all shell elements (see Figure 19.5). The composite was defined in ANSYS by

```
SECTYPE,1,SHELL
SECDATA,1.5,1,90      !# OF SECTION INTEGRATION POINTS PER
LAYER
SECDATA,1.5,1,0
SECDATA,1.5,1,90
SECDATA,1.5,1,0
SECDATA,1.5,1,90      !# OF SECTION INTEGRATION POINTS PER
LAYER
SECDATA,1.5,1,0
```

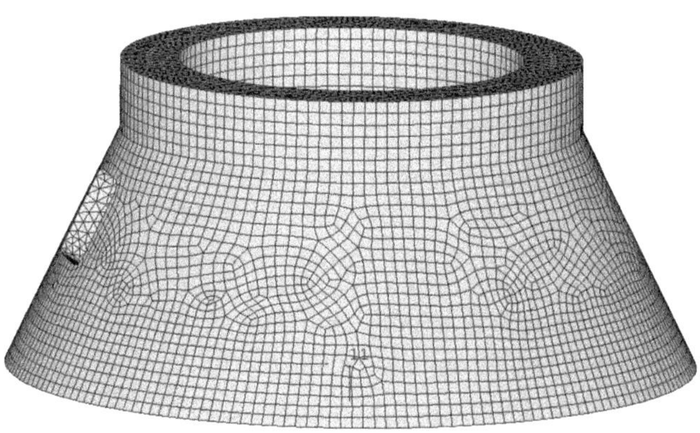

FIGURE 19.5 local coordinate system 11 defined for element coordinate system.

```
SECDATA,1.5,1,90
SECDATA,1.5,1,0
```

The parameters for damage evolution were specified in ANSYS by

```
TB,FCLI,1,1,,STRS  ! material strengths
TBDATA,1,0.283*6894   !FAILURE STRESS, X  DIRECTION TENSION
TBDATA,2,-0.215*6894  !FAILURE STRESS, X  DIRECTION COMPRESSION
TBDATA,3,6.96e-3*6894    !FAILURE STRESS, Y  DIRECTION TENSION
TBDATA,4,-2.90e-2*6894 !FAILURE STRESS, Y  DIRECTION
COMPRESSION
TBDATA,5,6.96e-3*6894    !FAILURE STRESS, Z  DIRECTION TENSION
TBDATA,6,-2.90e-2*6894 !FAILURE STRESS, Z  DIRECTION
COMPRESSION
TBDATA,7,1.15e-2*6894     !FAILURE STRESS, XY  DIRECTION
TENSION
TBDATA,8,7.25e-3*6894     !FAILURE STRESS, YZ  DIRECTION
TENSION
TBDATA,9,1.15e-2*6894     !FAILURE STRESS, XZ  DIRECTION
TENSION

TB,DMGI,1,1,,FCRT  ! FC for damage initiation
TBDATA,1,2,2,2,2  ! Max stress criteria for all four failure
modes

TB,DMGE,1,1,,MPDG  ! Damage evolution with MPDG method
TBDATA,1,0.6  ! 60% fiber tension damage (40% ultimate
strength)
TBDATA,2,0.6  ! 60% fiber compression damage (40% ultimate
strength)
TBDATA,3,0.6  ! 60% matrix tension damage (40% ultimate
strength)
TBDATA,4,0.6  ! 60% matrix compression damage (40% ultimate
strength)
```

19.3.3 Boundary Conditions and Loadings

The surface of the load head was loaded with a pressure of 240 MPa. The bottom surface of the conic part was fixed with all degrees of freedom (see Figure 19.6).

19.3.4 Results

The final deformation and stresses of the model with damage were plotted in Figure 19.7. For comparison, Figure 19.8 presents the final deformation and stresses of the model without damage. Both cases have very similar stress distributions because of the force loading. However, the mode with damage has a much larger deformation than that without damage. It was confirmed by the displacement history of one node located on the top surface (see Figure 19.9). It clearly shows that at the beginning, in the elastic stage, both models have an identical path. After the damage

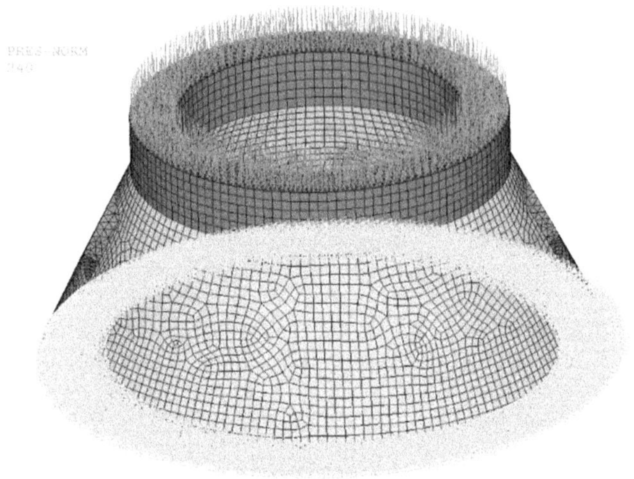

FIGURE 19.6 Boundary conditions and loading.

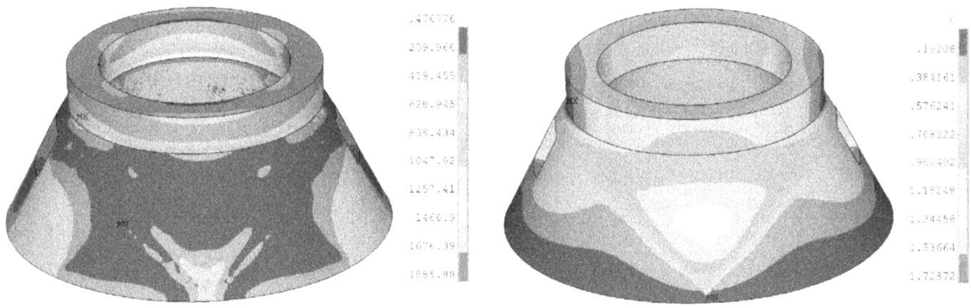

FIGURE 19.7 Final stresses and deformation of the model with damage. (a) von-Mises stresses (MPa). (b) Deformation (mm).

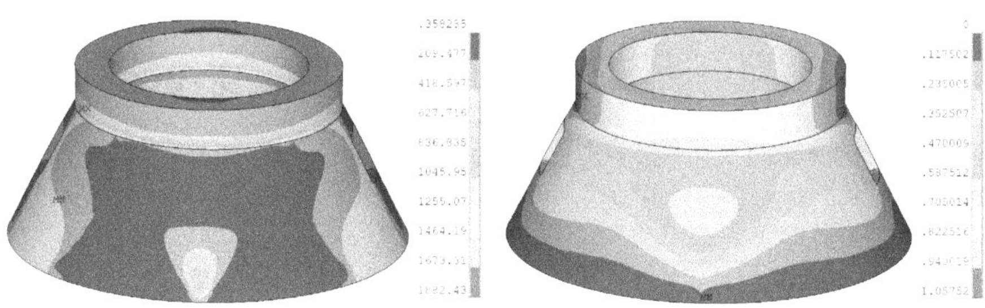

FIGURE 19.8 Final stresses and deformation of the model without damage. (a) von-Mises stresses. (MPa) (b) Deformation (mm).

initiates around loading 180 MPa, the displacement in the model with damage starts to shift from that in the model without damage, and their differences increase with the increase in loading.

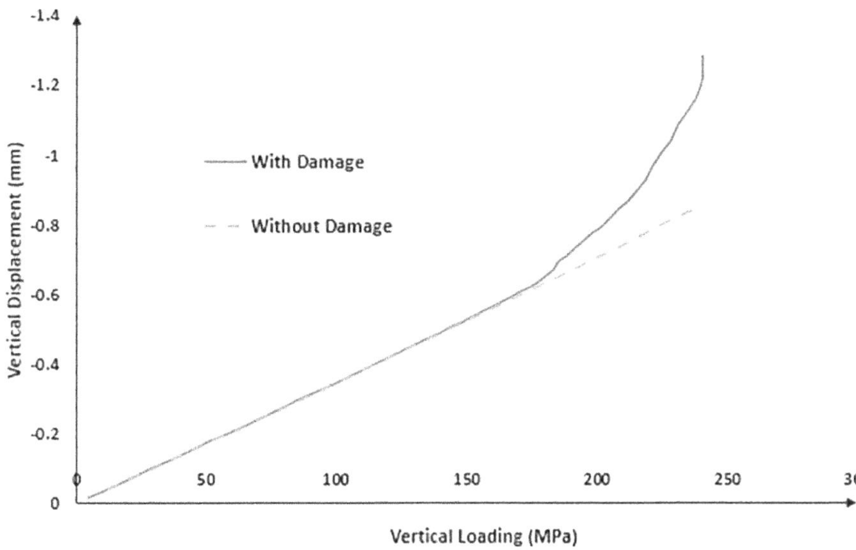

FIGURE 19.9 Displacement history of one node located on the top surface.

Figure 19.10 illustrates the damage to the composites on the outer and inner surfaces. The damage is initiated at the bottom and corners of the open doors and grows in two symmetrical branches. The damage starts slowly on the inner surface but grows faster on the outer surface. At the final stage, complete damage (marked as 2) occurs at the corners of the open doors on the inner surface.

19.3.5 Discussion

The damage process of the composite structure was simulated in ANSYS. Compared to the model without damage, the model with damage has much larger deformation under the same loading, and the differences between the two models increase with the increase in loading, which is close to reality.

For the solution part, the model with damage needs many iterations to reach convergence for each substep. On the other hand, the model without damage finishes the whole computation in a few steps.

The composite faces with 0° are in the axial direction. Thus, the local coordinate system with *X* in the vertical direction was defined as the element coordinate system. Therefore, the element *X* axis is determined from the projection of the local coordinate system on the shell element.

19.3.6 Summary

The composite in the composite structure failure test is modeled as a sandwich, and its damage was studied in ANSYS190. With the force loading, its deformation in the damage stage increases much faster than that of the model without damage.

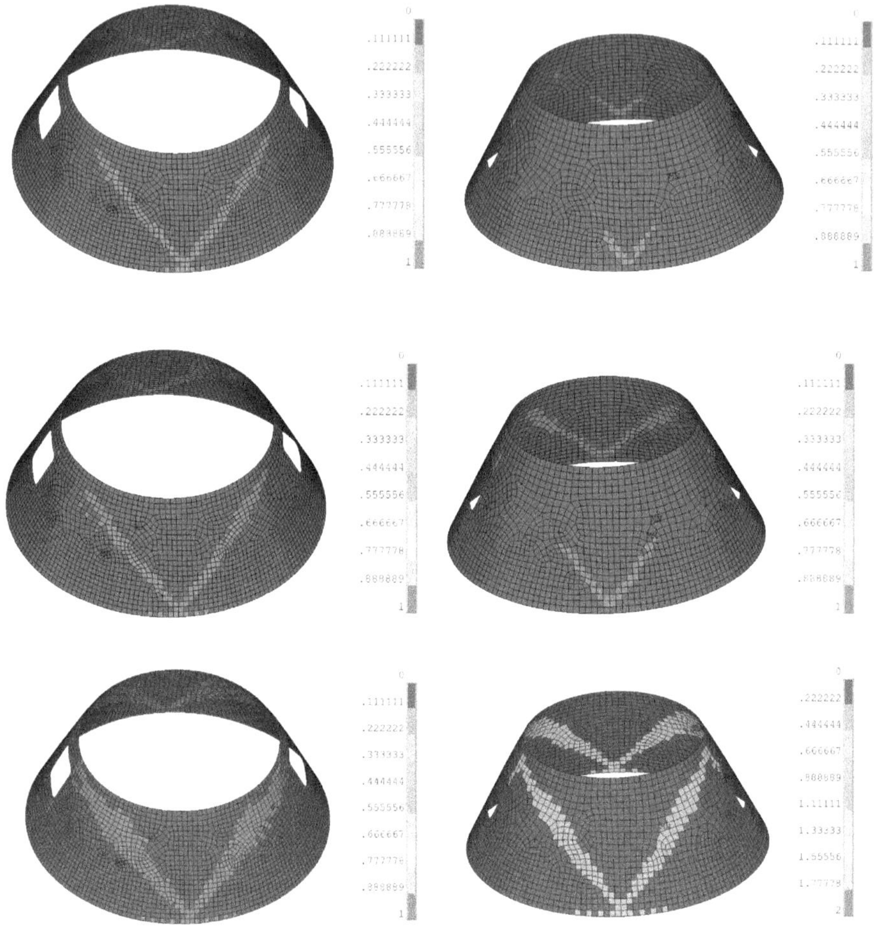

FIGURE 19.10 Damage of the composites in the outer and inner surfaces under different loading scales. (1a) outer side under loading 85%. (1b) inner side under loading 85%. (2a) outer side under loading 90%. (2b) inner side under loading 90%. (3a) outer side under loading 100%. (3b) inner side under loading 100%.

19.4 SIMULATION OF CRACK GROWTH IN SINGLE LEG BENDING PROBLEM

The composite structure has two issues that need to be addressed: (1) the damage, which was discussed in Section 19.3, and (2) delamination. The Virtual Crack Closure Technique (VCCT) has been developed to predict delamination in composite materials. In this section, the single leg bending (SLB) problem was studied using VCCT, which sets an example for the study of composite delamination.

19.4.1 Finite Element Model

One SLB problem was built in ANSYS [4] (see Figure 19.11). The whole model was meshed with PLANE182 (plane strain), while the interface was modeled

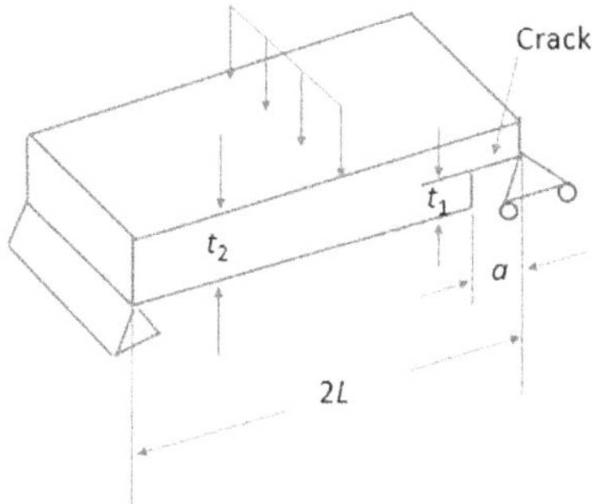

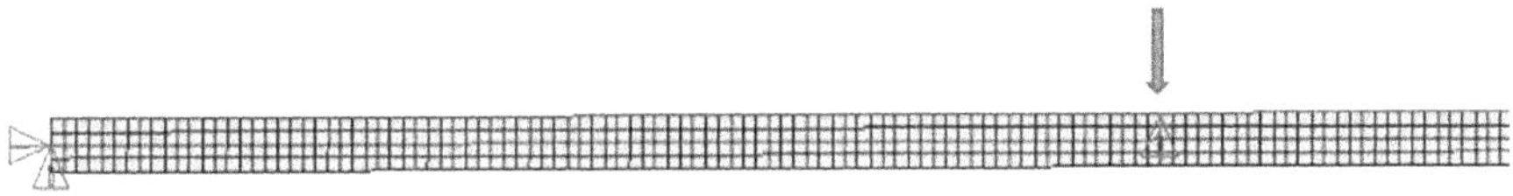

FIGURE 19.11 SLB problem and its finite element model (t_1 = 2.032 mm, t_2 = 4.064 mm, a = 34.3 mm, $2L$ = 177.8 mm). (a) SLB problem. (b) Finite element model.

with the Cohesive-Zone element (CZM)202 with plane strain. The initial crack was defined between the top area and bottom area.

19.4.2 Material Properties

The material of the composite was defined as anisotropic by

```
MP,EX,4,135.3E3              !* Material properties of lamina
MP,EY,4,9.0E3
MP,EZ,4,9.0E3
MP,GXY,4,4.5E3
MP,GYZ,4,4.5E3
MP,GXZ,4,3.3E3
MP,PRXY,4,0.24
MP,PRXZ,4,0.24
MP,PRYZ,4,0.46
```

A linear fracture criterion was specified for the crack growth with an energy release rate of 0.05 for critical modes I, II, and III as

```
TB,CGCR,1,,3,LINEAR            !* LINEAR FRACTURE CRITERION
TBDATA,1,0.05,0.05,0.05
```

19.4.3 Crack Definition

The crack was defined in the model using the CINT command as

```
CINT,NEW,1
CINT,TYPE,VCCT
CINT,CTNC,CRACK1
CINT,NORM,12,2
```

```
CINT,NCON,6            ! NUMBER OF CONTOURS
CINT,SYMM,OFF          ! SYMMETRY OFF
```

And the crack growth was specified as

```
CGROW,NEW,1
CGROW,METHOD,VCCT
CGROW,CID,1           ! CINT ID  VCCT CALCULATION
CGROW,CPATH,CPATH1     ! CRACK PATH
```

19.4.4 Boundary Conditions and Loadings

The center of the left end was fixed in the vertical direction, and the right end was constrained with all degrees of freedom. The top center was loaded with a displacement of 5 mm (see Figure 19.11).

19.4.5 Results

The final deformation of the model was illustrated in Figure 19.12. It clearly shows that a crack grows while the model is bending. The length of crack growth 82.9 mm was obtained using the command PRCI,1,CEXT in the postprocess. Command PRCINT,1,VCCT prints out the energy release rate of the last step,

$$G1 = 0.0274 \text{ MPa} \cdot \text{mm}$$

$$G2 = 0.0233 \text{ MPa} \cdot \text{mm}$$

$$G3 = 0.0 \text{ MPa} \cdot \text{mm}$$

$$Gtot = 0.0507 \text{ MPa} \cdot \text{mm}$$

Obviously, the energy release rates of modes I and II are very close, while they are zero for mode III.

Figure 19.13 plots the reaction force with the displacement loading. At the starting stage, the reaction force increases with the displacement loading. After the crack growth starts, the reaction force drops significantly and remains constant while the crack growth continues. When the crack growth stops, the reaction force increases linearly with an increase in the displacement loading.

19.4.6 Discussion

Different from the stress intensity factor (K) and J-integral in fracture mechanics that are suitable for isotropic materials, VCCT emphasizes energy release rate and can

FIGURE 19.12 Deformation of the model (Dmax = 10 mm).

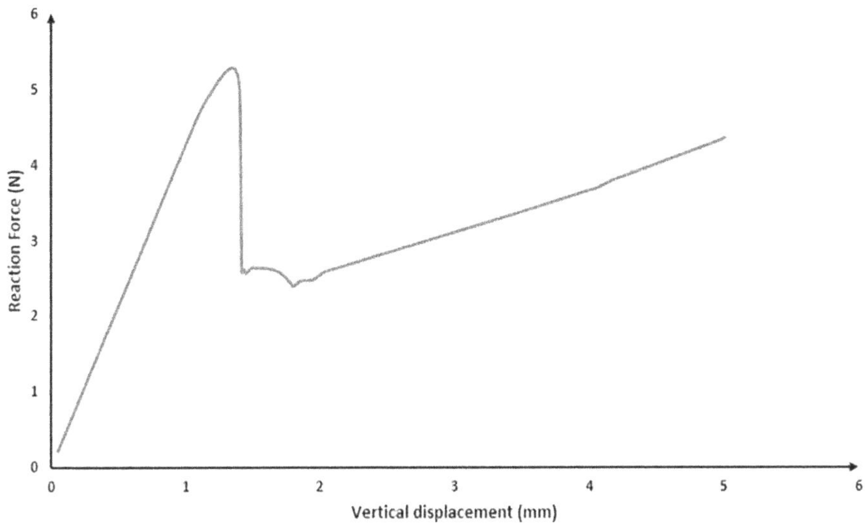

FIGURE 19.13 Reaction force with the displacement loading.

be used to model delamination in composite materials. The crack growth of the SLB problem was simulated using VCCT in ANSYS190. Unlike the double cantilever beam (DCB) problem, in which mode I dominates, both modes I and II play a close role in the crack growth of SLB.

ANSYS supports the cohesive-zone model (CZM) method and the Virtual Crack Closure Technique (VCCT) method. The SLB problem can be studied using the CZM method as well.

19.4.7 Summary

The SLB was modeled in ANSYS190, and its crack growth was simulated using VCCT. The results indicate that both mode I and mode II are involved in the crack growth.

REFERENCES

[1] Jones, R.M., *Mechanics of Composite Materials*, CRC Press, New York, 1998.
[2] ANSYS190 Help Documentation in the help page of product ANSYS190.
[3] Frappier, A.M., An investigation of composite failure analyses and damage evolution in finite element models, Purdue University, Master Theses, 2006.
[4] Krueger, R., Shivakumar, K., and Raju, I.S., Fracture mechanics analyses for interface crack problems - a review, *54th AIAA/ASME/ASCE/AHS/ASC Structures, Structural Dynamics, and Materials Conference, Structures, Structural Dynamics, and Materials and Co-located Conferences*, Boston, Massachusetts, 2013.

20 Functionally Graded Material

Functionally Graded Materials (FGMs), such as bones and teeth, exist in the human body; these materials are designed to meet function requirements. FGMs are derived from nature to solve engineering problems. Different kinds of fabrication processes are developed to produce FGMs. Chapter 20 discusses functionally graded materials, including their introduction in Section 20.1, their modeling using ANSYS in Section 20.2, and one example of an FGM in Section 20.3.

20.1 INTRODUCTION OF FUNCTIONALLY GRADED MATERIAL

As revolutionary materials, FGMs are a set of advanced materials with changing properties over a varying dimension. The difference between FGMs and composite materials is that sharp interfaces exist in the composite, which cause fracture and failure. On the other hand, FGMs have a smooth transition from one material to another by using a gradient interface.

Generally, FGMs are classified into two groups: (1) thin FGMs with a relatively thin surface coating and (2) bulk FGMs with volumes of materials requiring more intensive processes [1]. Thin section or surface coating FGMs are fabricated using physical or chemical vapor deposition (PVD/CVD), plasma spraying, or self-propagating high-temperature synthesis (SHS), while bulk FGMs are produced using powder metallurgy, centrifugal casting method, and solid freeform technology.

With improvements in fabrication processes, the overall process cost for manufacturing FGMs has been dramatically reduced. Thus, the application of FGMs has been expanded in many fields, such as aerospace, automobile industry, medicine, sports, energy, sensors, and optoelectronics.

20.2 MATERIAL MODEL OF FUNCTIONALLY GRADED MATERIAL

The material properties of FGMs are a function of location, which can be modeled in ANSYS through two different approaches. The first approach, the interpolation algorithm, uses TBIN, ALGO to define the material properties as

```
TB,ELAS,1        ! DEFINES THE MATERIAL PROPERTIES
*DO, I, NP       ! NP IS THE NUMBER OF SUPPORTING POINTS
! ASSIGN YOUNG'S MODULUS ENP FOR THE SUPPORTING POINTS
  E%I%= ENP(I)
! (XXI, YYI) IS THE LOCATION OF SUPPORTING POINT I
  TBFIELD, XCOR,     XX%I%
```

DOI: 10.1201/9781003436317-24

```
  TBFIELD, YCOR,      YY%I%
  TBDATA,   1,        E%I% ,   NU  ! NU IS POISSON'S RATIO
*ENDDO
TBIN,ALGO
```

In the second approach, every element has its own material identity, with material parameters computed from the location of the element center. The approach is implemented by DO LOOP:

```
! DEFINE THE MATERIAL PROPERTIES
*DO, I, 1, NE        ! NE IS THE NUMBER OF ELEMENTS
  E%I%=EP            ! EP IS THE YOUNG'S MODULUS OF ELEMENT I
  MP,EX,I,E%I%
  MP,NUXY,I,NU       ! NU IS POISSON'S RATIO OF ELEMENT I
*ENDDO
```

The difference between the above approaches is that the first approach has only one material identity, while the second approach has as many material identities as the number of elements. Therefore, the first approach has an advantage; it is insensitive to meshing, while the second approach is sensitive to meshing. With finer meshing and more elements, the second approach gets closer to the first approach.

20.3 SIMULATION OF A SPUR GEAR WITH FUNCTIONALLY GRADED MATERIALS

Spur gears have a wide application in power transmission. The hardworking conditions of the spur gear require them to resist tooth deflection and stress concentration, which demand that the material of spur gear has sufficient strength. A FGM is a revolutionary material and was selected for fabricating a spur gear in this study; the gear was simulated using two different methods. The results can be used as a guideline for the mechanical design of spur gears.

20.3.1 Finite Element Model

A spur gear was meshed with PLANE182, assuming plane stress (Figure 20.1) [2].

20.3.2 Material Properties

The spur gear was assumed to be made of a FGM, following the exponential law [2,3]

$$E = 2.01 \times 10^5 e^{0.0097(x-14.9)} \tag{20.1}$$

The material properties of the spur gear were defined in two different ways. The first one was the interpolation algorithm using the command TBIN, ALGO. The material of the gear was defined as

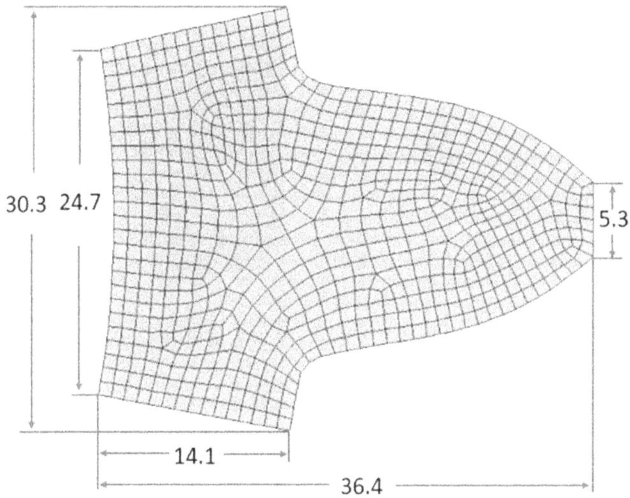

FIGURE 20.1 Finite element model of a spur gear (all dimensions in mm).

```
TB,ELAS,1                    ! DEFINE THE MATERIAL PROPERTIES
*DO, I, 1, 11
! ASSIGN YOUNG'S MODULUS FOR THE SUPPORTING POINTS
E%I%=2.01E5*EXP(0.0097*(I-1)*2)
TBFIELD, XCOR,      (I-1)*2+14.9
TBDATA,   1,          E%I% ,   0.0
*ENDDO
```

The second approach defined a number of material models, which have individual Young's moduli calculated by equation (20.1) using different x values. These material models were assigned to the appropriate area of the spur gear according to the corresponding x value.

```
! DEFINE THE MATERIAL PROPERTIES
*DO, I, 1, 6
! ASSIGN YOUNG'S MODULUS FOR THE ELEMENTS
E%I%=2.01E5*EXP(0.0097*(I-1)*4)
MP,EX,I,E%I%
MP,NUXY,I,0.
*ENDDO
```

Both approaches have a very close distribution of Young's modulus (see Figure 20.2).

20.3.3 Loadings and Boundary Conditions

The inner rim of the gear tooth and the two radial lines were constrained in both the x and y directions. A tangential force of 113.68 N was applied at the tip of the gear tooth [2]. To reduce stress concentration, the force was loaded equally on two nodes (see Figure 20.3).

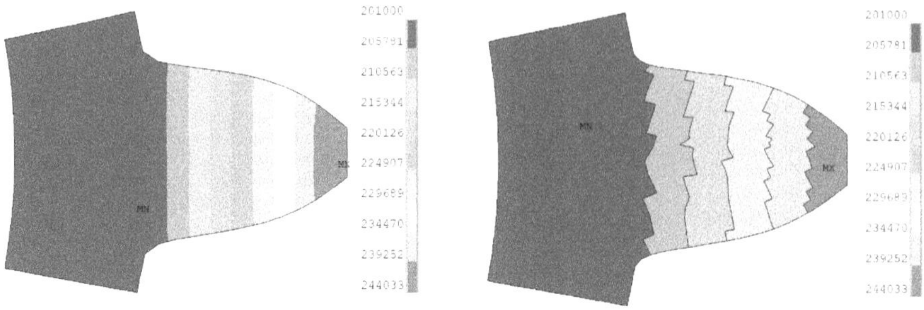

FIGURE 20.2 Young's modulus distribution of the spur gear through. (a) TBField. (b) Equivalent materials.

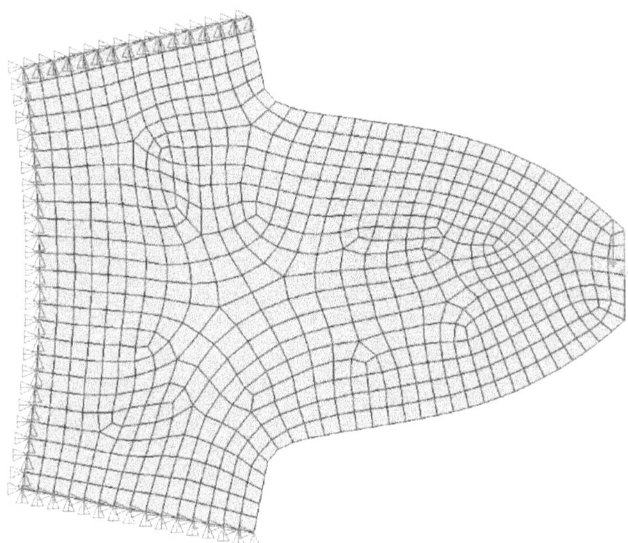

FIGURE 20.3 Boundary conditions and loadings.

20.3.4 Results

Figure 20.4 depicts the deformation of the gear. Both approaches result in almost the same deflection, about 8.5e-5 mm. The von-Mises stresses of the gear are plotted in Figure 20.5. The maximum stress occurs at the loading area due to stress concentration, about 0.92 MPa. The stress results of the two approaches are very close.

20.3.5 Discussions

The spur gear made using a FGM was simulated using two different methods. Both methods resulted in very close displacements and stresses, which validate the interpolation algorithm of ANSYS.

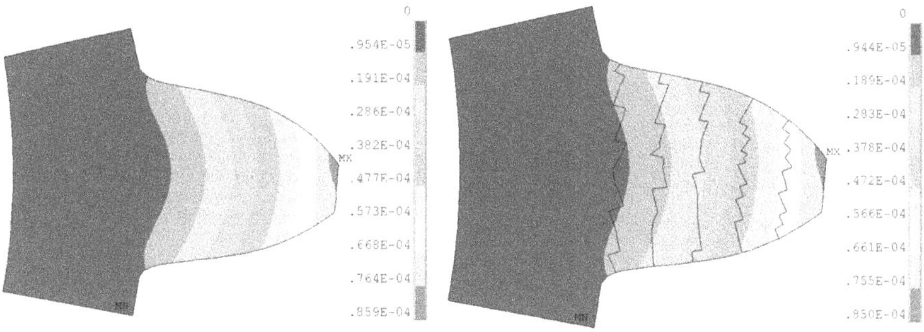

FIGURE 20.4 Deformation of the spur gear (mm). (a) TBField. (b) Equivalent materials.

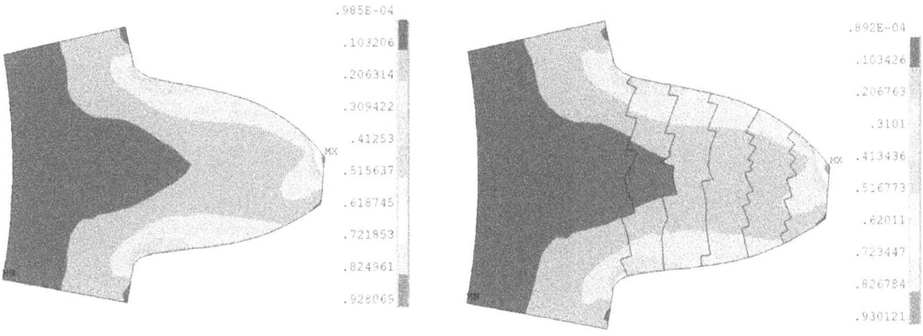

FIGURE 20.5 von-Mises stresses of the spur gear (MPa). (a) TBField. (b) Equivalent materials.

The gear defined by TBIN, ALGO has a continuous Young's modulus with one material identity. On the contrary, Young's modulus obtained by the equivalent material model is discontinuous and defined by a number of material identities.

20.3.6 Summary

The spur gear was modeled as FGM by two different methods – the interpolation algorithm and the equivalent material model. Both showed very close results in terms of deformation and stress distribution. Unlike the equivalent material model, the interpolation algorithm defined the FGM with one material identity and showed a continuous distribution of Young's modulus.

REFERENCES

[1] Mahamood, R.M., Akinlabi, E. T., Shukla, M., and Pityana, S., Functionally graded material: an overview, *Proceedings of the World Congress on Engineering 2012 Vol III, WCE2012*, London, UK, July, 2012.

[2] Kumar., R., and Sarda, A.A., Analysis of functionally graded material spur gear under static loading condition, *Advance Physics Letter*, Vol. 3, 2016, pp. 1–6.

[3] Pawar, P.B., and Utpat, A.A., Analysis of composite material spur gear under static loading condition, *Materials Today: Proceedings*, Vol. 2, 2015, pp. 2968–2974.

21 Shape Memory Alloys

Shape memory alloys (SMAs) refer to a special type of metal that can recover its initial shape after severe deformations. Their unique features make them widely applicable in industry-for orthodontic wires, eyeglasses, and stents in the biomedical area; for actuators and connectors in the mechanical area; and for shape and vibration in the structural area. In Chapter 21, the structure of SMAs and the corresponding material models are introduced, followed by SMAs' 2019 applications for an orthodontic wire and a vacuum-tight shape memory flange.

21.1 STRUCTURE OF SMA AND VARIOUS MATERIAL MODELS

21.1.1 Structure of SMA

SMAs have two different phases: (1) the high temperature phase named austenite (A) and (2) the low temperature one called martensite (M). Both have different crystal structures and, therefore, different properties [1]. Figure 21.1 illustrates the crystal structure of austenite, which is generally cubic. On the other hand, martensite exists in two forms: twinned martensite (see Figure 21.2a) and detwinned martensite (see Figure 21.2b). The phase transformation from austenite to martensite and vice versa generates the unique features of SMAs, which are stated as follows.

21.1.1.1 Superelasticity

The superelasticity of SMAs relates to stress-induced transformation at a temperature above A_f. Consider a loading path ($A \rightarrow B \rightarrow C \rightarrow D \rightarrow E \rightarrow F \rightarrow A$) as shown in Figure 21.3. When a mechanical load is applied up to B, SMA has an elastic deformation in the austenite phase. After the load reaches B (σ^{Ms}) (see Figure 21.4), a phase transformation starts accompanying the transformation strain. The transformation continues until C (σ^{Mf}), which is the end of the transformation. From C to D, SMA has an elastic deformation in the detwinned martensite phase. In the unloading stage of $D \rightarrow E$,

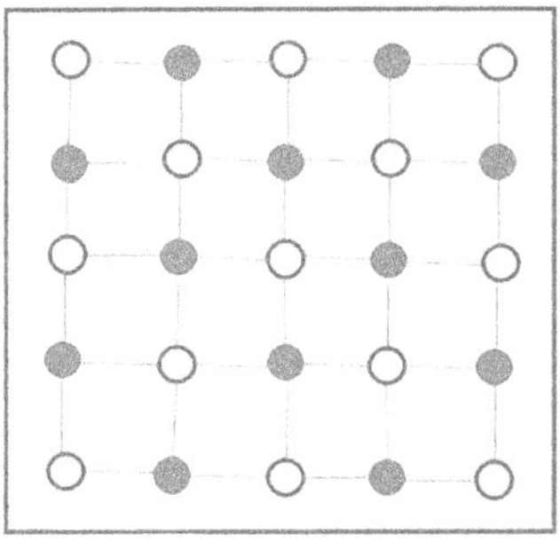

FIGURE 21.1 Crystal structure of austenite.

DOI: 10.1201/9781003436317-25

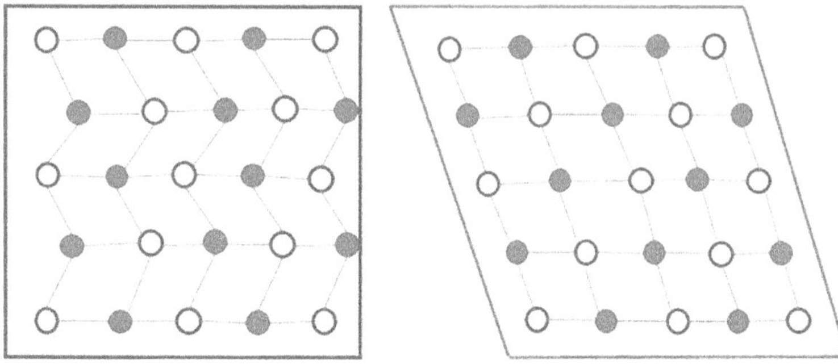

FIGURE 21.2 Structure of martensite: (a) twinned martensite and (b) detwinned martensite.

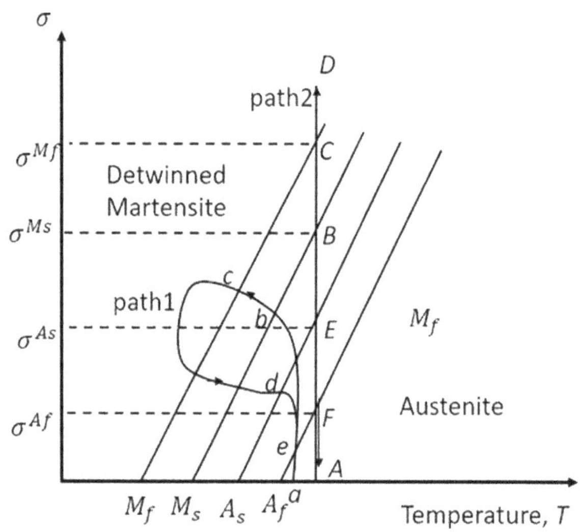

FIGURE 21.3 Stresses of SMAs with temperature [1].

the SMA remains elastic. At point E, the stress reaches σ^{As} that causes the martensite to return to austenite. The whole phase transformation occurs between E and F. F is the end of phase transformation with stress σ^{Af}. At the final unloading stage ($F \rightarrow A$), the SMA returns to its origin elastically.

Obviously, the superelasticity of SMAs is much different from the large deformation of elastic materials. For elastic materials, the unloading path always retraces the loading path. However, the SMAs' 2019 unloading path differs from their loading path because σ^{Mf} is different from σ^{As}, and σ^{Ms} is not equal to σ^{Af}, which causes the phase transformation to occur at different stress levels.

21.1.1.2 Shape Memory Effect

Assume a loading path ($A \rightarrow B \rightarrow C \rightarrow D \rightarrow E \rightarrow A$) as shown in Figure 21.5. $A \rightarrow B$ is a cooling stage. When the temperature drops below the forward transformation

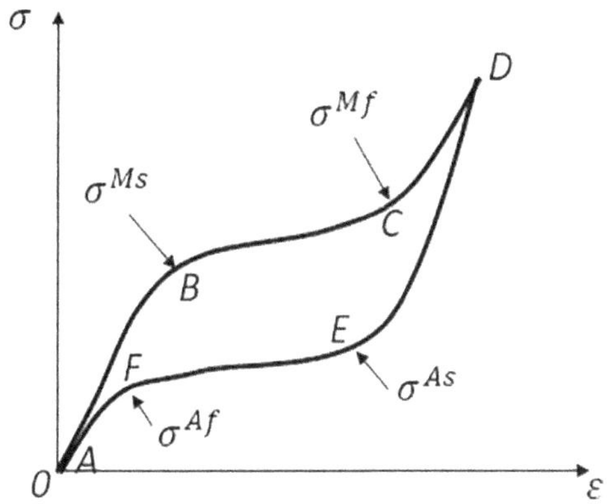

FIGURE 21.4 Various stresses of SMAs with strains.

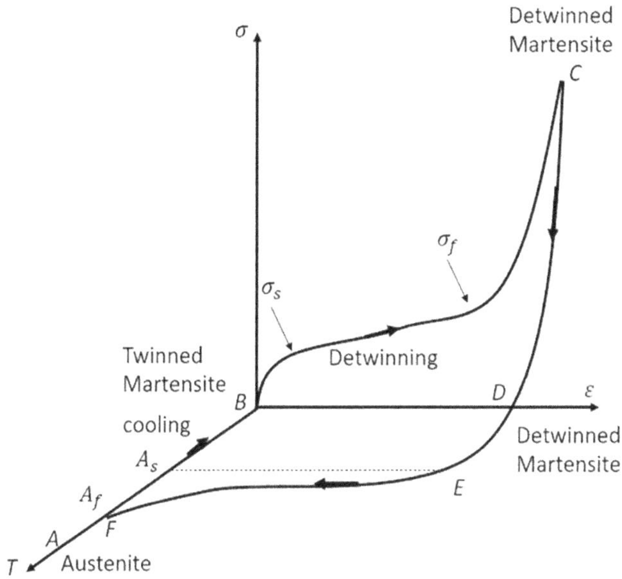

FIGURE 21.5 Shape memory effect [1].

temperature (M_s and M_f), the phase of SMA changes from the parent phase to the twinned martensite. In the loading stage from B to C, when the stress state of SMA is above the starting stress level σ_s, the reorientation starts. The detwinning process continues with an increase in loading and ends at a stress level σ_f. Detwinned martensite remains and deforms elastically in the unloading stage from C to E. The heating process is $D \rightarrow E \rightarrow F$. When the temperature reaches A_s (point E), the transformation from detwinned martensite to austenite starts and then completes at A_f (point F). Thus, the original shape of the SMA is regained.

21.1.2 Various SMA Material Models

The aforementioned SMA features can be modeled in ANSYS by using two different material models [2]: SMA model for superelasticity and SMA material model with the shape memory effect.

21.1.2.1 SMA Model for Superelasticity

The elastic strain ε_e is obtained by subtracting the transformation strain ε_{tr} from the total strain ϵ:

$$\varepsilon_e = \varepsilon - \varepsilon_{tr} \tag{21.1}$$

Thus, the stress σ is calculated by

$$\sigma = \mathbf{D} : \varepsilon_e = \mathbf{D} : (\varepsilon - \varepsilon_{tr}) \tag{21.2}$$

The Drucker-Prager model is used to describe the phase transformation:

$$F = q + 3\alpha \cdot p \tag{21.3}$$

where

$$q = \sqrt{\frac{3}{2}\mathbf{S} : \mathbf{S}} \tag{21.4}$$

$$\mathbf{S} = \sigma - p1 \tag{21.5}$$

$$p = \frac{1}{3}\sigma : 1 \tag{21.6}$$

α is the material parameter that takes into account the different material behaviors of tension and compression.

Thus, the evolution of the martensite fraction, ξ_s, is a function of F

$$\dot{\xi}_s = \left\{ \begin{array}{ll} -H^{AS}(1-\xi_s)\dfrac{\dot{F}}{F - R_f^{AS}} & A \rightarrow S \\ H^{SA}\xi_s \dfrac{\dot{F}}{F - R_f^{SA}} & S \rightarrow A \end{array} \right\} \tag{21.7}$$

where

$$R_f^{AS} = \sigma_f^{AS}(1+\alpha) \tag{21.8}$$

$$R_f^{SA} = \sigma_f^{SA}(1+\alpha) \tag{21.9}$$

$$H^{SA} = \begin{cases} 1 & \text{if } R_s^{SA} < F < R_f^{SA} \quad \text{and } \dot{F} > 0 \\ 0 & \text{otherwise} \end{cases} \tag{21.10}$$

$$H^{SA} = \begin{cases} 1 & \text{if } R_f^{SA} < F < R_s^{SA} \quad \text{and } \dot{F} < 0 \\ 0 & \text{otherwise} \end{cases} \tag{21.11}$$

$$R_s^{AS} = \sigma_s^{AS}(1+\alpha) \tag{21.12}$$

$$R_s^{SA} = \sigma_s^{SA}(1+\alpha) \tag{21.13}$$

Therefore, the transformation strain tensor, ε_{tr}, is expressed in terms of ξ_s and yield function F:

$$\dot{\varepsilon}_{tr} = \dot{\xi}_s \bar{\varepsilon}_L \frac{\partial F}{\partial \sigma} \tag{21.14}$$

$\bar{\epsilon}_L$ is the maximum residual strain.

Equations (21.1–21.14) include these material parameters: σ_s^{AS}, σ_f^{AS}, σ_s^{SA}, σ_f^{SA}, $\bar{\varepsilon}_L$, and α. Therefore, the SMA model for superelasticity is defined in ANSYS by commands:

```
TB, SMA,,,,SUPE
TBDATA, 1, σsAS, σfAS, σsSA, σfSA, ε̄L, α
```

21.1.2.2 SMA Model with Shape Memory Effort

Similar to the SMA with superelasticity, the total stress in the SMA with the shape memory effect is calculated by

$$\boldsymbol{\sigma} = \mathbf{D} : (\boldsymbol{\varepsilon} - \boldsymbol{\varepsilon}_{tr}) \tag{21.15}$$

where $\mathbf{D}$ is a linear function of ε_{tr}

$$\mathbf{D} = \frac{\|\boldsymbol{\varepsilon}_{tr}\|}{\boldsymbol{\varepsilon}_L}(\mathbf{D}_M - \mathbf{D}_A) + \mathbf{D}_A \tag{21.16}$$

When the material is in its austenite phase, $\mathbf{D} = \mathbf{D}_A$; when the material is in its martensite phase, $\mathbf{D} = \mathbf{D}_M$.

A Prager-Lode type is selected for the yield function to take into account the asymmetric behavior of the SMA under tension and compression:

$$F(\mathbf{X}_{tr}) = \sqrt{2J_2} + m\frac{J_3}{J_2} - R \tag{21.17}$$

where

$$J_2 = \frac{1}{2}\left(\mathbf{X}_{tr}^2 : 1\right) \tag{21.18}$$

$$J_3 = \frac{1}{3}\left(\mathbf{X}_{tr}^3 : 1\right) \tag{21.19}$$

$$\mathbf{X}_{\mathbf{tr}} = \boldsymbol{\sigma}' - \left[\tau_M(T) + h\|\boldsymbol{\varepsilon}_{tr}\| + \gamma\right]\frac{\boldsymbol{\varepsilon}_{\mathbf{tr}}}{\|\boldsymbol{\varepsilon}_{\mathbf{tr}}\|} \tag{21.20}$$

$$\tau_M(T) = \left\langle \beta(T - T_0)\right\rangle^+ \tag{21.21}$$

β is the temperature scaling parameter;
T is the working temperature;
T_0 is the temperature M_f below which no twinned martensite occurs;
h is the hardening parameter; and
γ is given by

$$\left\{ \begin{array}{ll} \gamma = 0 & \textit{if } 0 < \|\boldsymbol{\varepsilon}_{tr}\| < \varepsilon_L \\ \gamma \geq 0 & \textit{if } \|\boldsymbol{\varepsilon}_{tr}\| = \varepsilon_L \end{array} \right\} \tag{21.22}$$

ε_L is the maximum transformation strain;
m is the Lode dependency parameter; and
R is the elastic limit.

The evolution of $\boldsymbol{\varepsilon}_{tr}$ is related with the yield function F,

$$\boldsymbol{\varepsilon}_{\mathbf{tr}} = \boldsymbol{\varepsilon}_{\mathbf{tr}}(n) + \Delta\xi\frac{\partial \boldsymbol{F}}{\partial \sigma} \tag{21.23}$$

Equations (21.15–21.23) cover seven material parameters, including h, T_0, R, β, ε_L, Em, and m. Thus, the SMA model with the shape memory effect is defined in ANSYS by the commands:

```
TB,SMA,,,,MEFF
TBDATA, 1, h, T0, R, β, εL
TBDATA,6, Em, m
```

21.1.3 Definition of Material Parameters

21.1.3.1 SMA with Superelasticity

Some material parameters of SMAs with superelasticity, such as E, σ_s^{AS}, σ_f^{AS}, σ_s^{SA}, σ_f^{SA}, and $\bar{\varepsilon}_L$, can be determined by the stress-strain curve of the uniaxial tension test illustrated in Figure 21.6. The material parameter α specifies the material response under tension and compression. If tension and compression have the same response, then $\alpha = 0$. Generally, α is defined by the starting stress of the austenite to martensite phase transformation under tension and compression as

$$\alpha = \frac{\sigma_c^{AS} - \sigma_t^{AS}}{\sigma_c^{AS} + \sigma_t^{AS}} \tag{21.24}$$

where σ_c^{AS} is the initial stress of austenite to martensite phase transformation under compression.

σ_t^{AS} is the initial stress of austenite to martensite phase transformation under tension.

21.1.3.2 SMA with Shape Memory Effect

E_A and E_m are determined by the slope of the elastic stages in the SMA stress-strain curve (see Figure 21.7). H is the slope of the transformation strain-stress curve, and ϵ_L is the maximum transformation strain (see Figure 21.8). T_0 is M_f, and β is the slope of the temperature-stress plot of SMA (see Figure 21.9).

R and m are determined by tension limit σ_t and compression limit σ_c by

$$R = 2\sqrt{2/3}\,\frac{\sigma_c \sigma_t}{\sigma_c + \sigma_t} \tag{21.25}$$

$$m = \sqrt{27/2}\,\frac{\sigma_c - \sigma_t}{\sigma_c + \sigma_t} \tag{21.26}$$

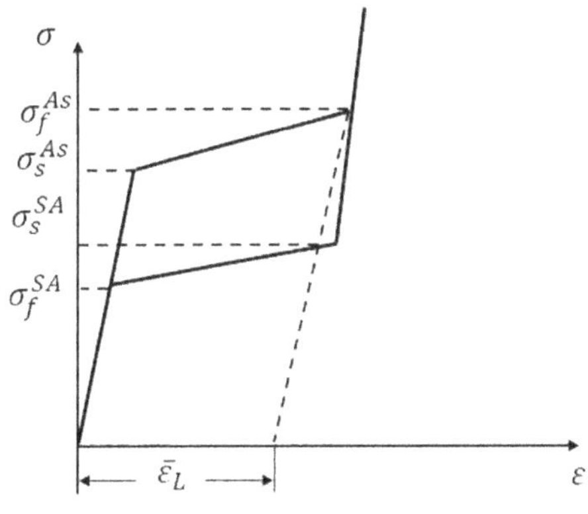

FIGURE 21.6 Stress-strain of SMA with superelasticity.

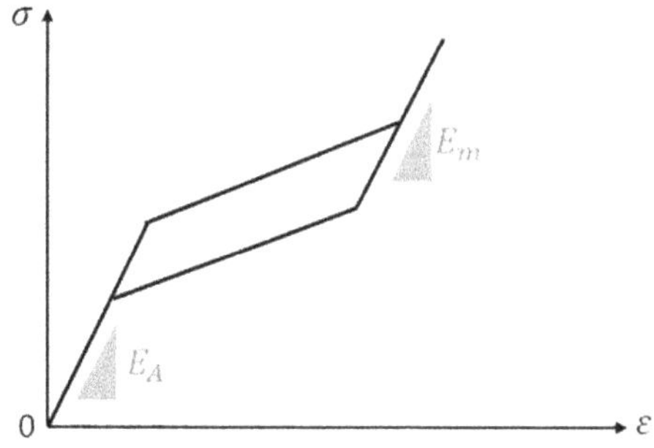

FIGURE 21.7 Stress-strain curve of SMAs.

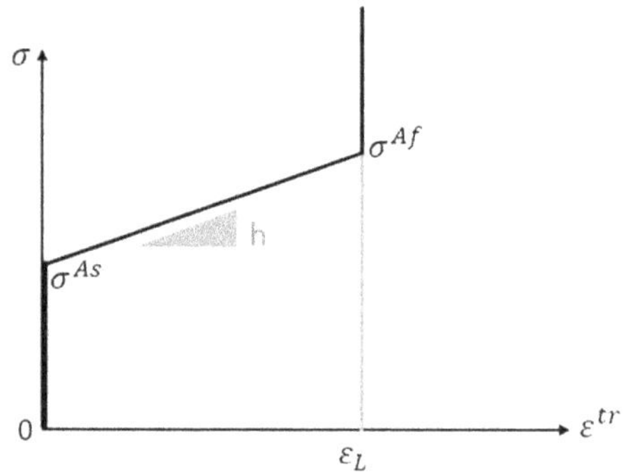

FIGURE 21.8 Stress-transformation strain curve of SMAs.

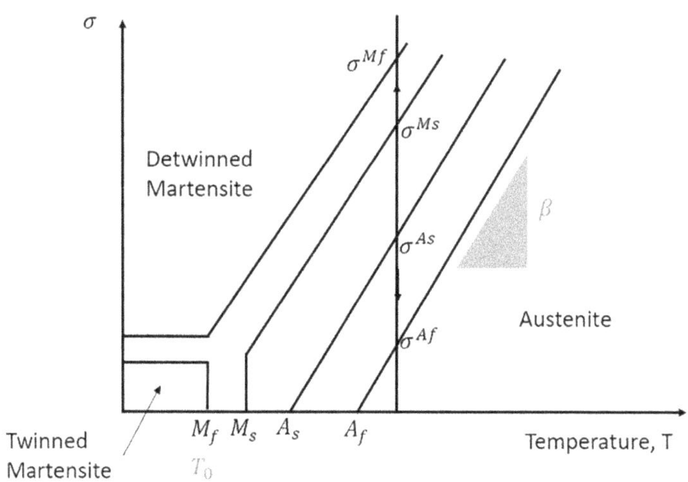

FIGURE 21.9 Temperature-stress curve of SMAs [1].

21.2 SIMULATION OF ORTHODONTIC WIRE

Biocompatibility studies indicate that SMA NiTi alloys have low cytotoxicity and genotoxicity [3,4], which allows for the possibility of their application in orthodontics. The orthodontic wire (see Figure 21.10), as a part of the fixed orthodontic appliance, delivers a force on teeth that permits tooth movement without tissue damage. In the early stages of orthodontic treatment, wires made using SMA NiTi alloys are used for tooth leveling and alignment. In this study, a wire made using SMA NiTi alloys was simulated using the ANSYS SMA model for superelasticity.

21.2.1 Finite Element Model

Because of its symmetry, half of the finite element model was developed in ANSYS190 (see Figure 21.11) [5]. The model was meshed using SOLID185 with 1,856 elements and 2,925 nodes.

21.2.2 Material Properties

The wire was modeled using the SMA material model with superelasticity. The material parameters are listed in Table 21.1 [5].

21.2.3 Loadings and Boundary Conditions

The whole simulation was completed with two loading steps. In the first step, the right end was pulled in the *x* direction by 7 mm. Then, it was brought back to its initial position in the second step.

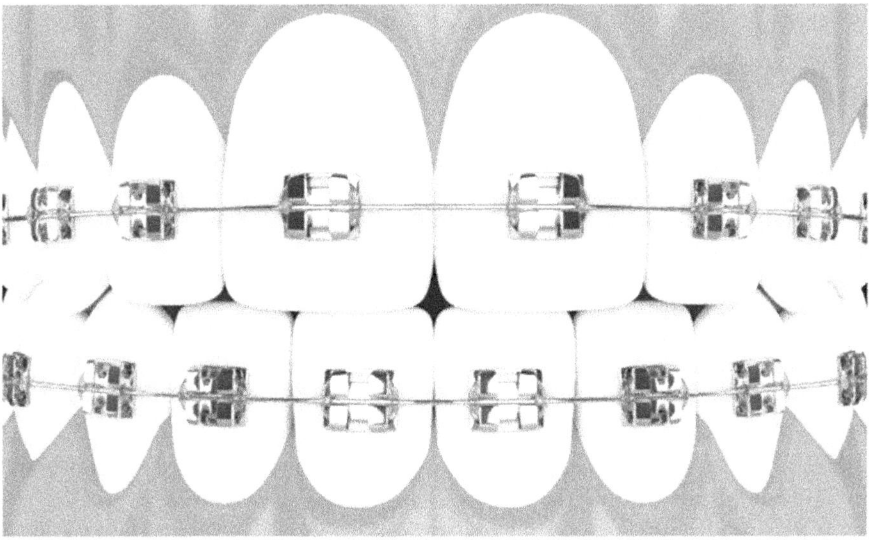

FIGURE 21.10 Orthodontic wires. (Aleksandra Gigowska ©123RF.com.)

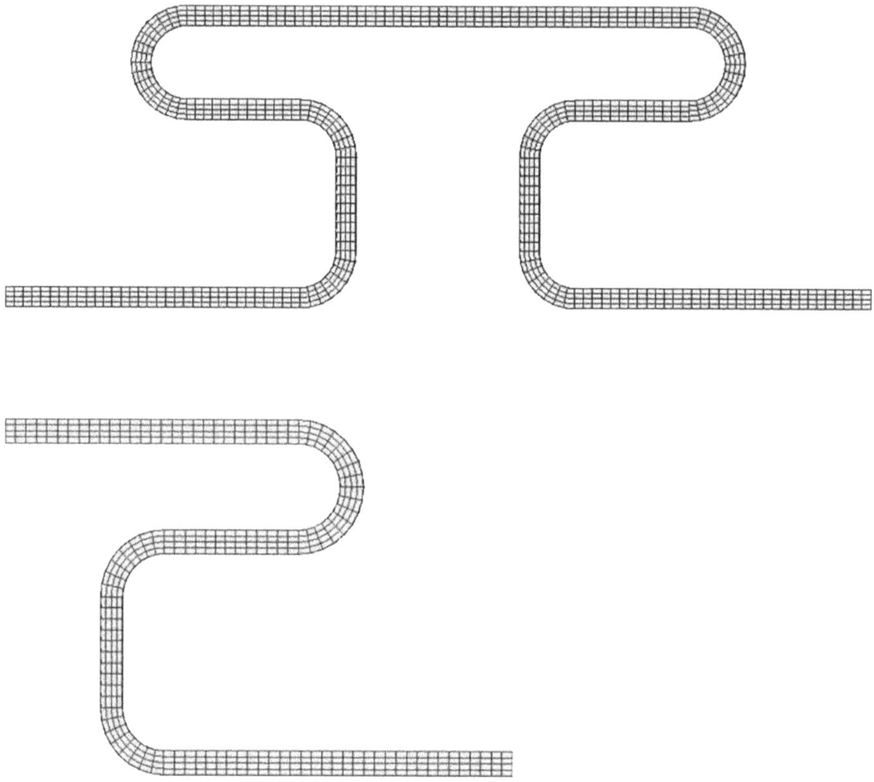

FIGURE 21.11 Finite element model of a wire. (a) Full model and (b) half model with symmetrical conditions.

TABLE 21.1
Material Parameters of Wire

E_A (MPa)	v	σ_s^{AS} (MPa)	σ_f^{AS} (MPa)	σ_s^{SA} (MPa)	σ_f^{SA} (MPa)	$\bar{\epsilon}_L$	α
50,000	0.36	500	500	300	300	0.07	0.136

The symmetric surface was constrained with all degrees of freedom as boundary conditions (see Figure 21.12).

21.2.4 Results

Figures 21.13–21.15 plot the deformation, von-Mises stresses, and transformation strains of the first step, respectively. The deformation and the von-Mises stresses of the second step are presented in Figures 21.16 and 21.17, respectively.

At the first step, the wire has a maximum deformation of 8.1 mm. The maximum von-Mises stress of 927.9 MPa and the maximum transformation strain of 0.061

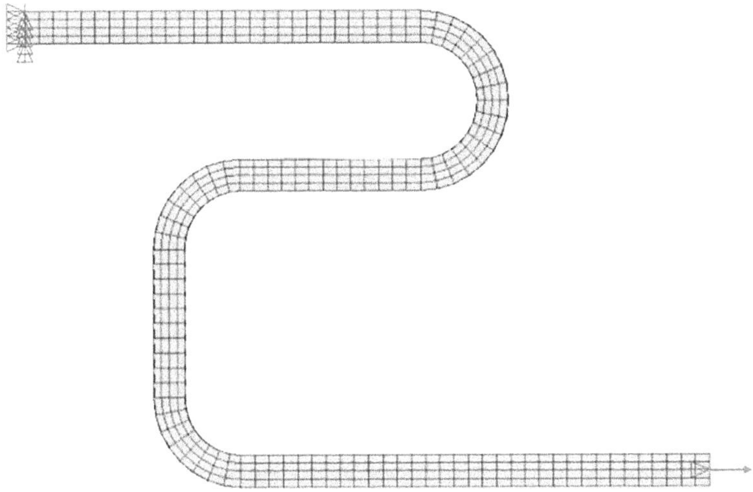

FIGURE 21.12 Boundary condition of orthodontic wire.

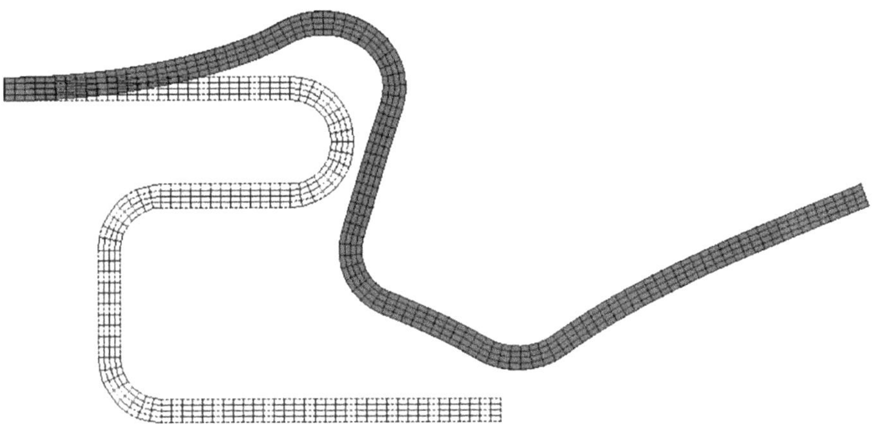

FIGURE 21.13 Deformation of the orthodontic wire at the end of loading stage (DMX = 8.1 mm).

occur at the same location. In the second step, after the wire returns to its initial position, the deformation reduces to near zero, and the maximum von-Mises stress becomes as low as 3.54 MPa, which validates the superelasticity of the SMA.

Figure 21.18 illustrates the von-Mises stresses and plastic strains of element A located at the maximum stress area with time. At the beginning of the first step, the stress increases linearly without phase transformation. After it reaches phase transformation, the transformation strain increases quickly, while the stress increases flatly. In the early stages of the second step, the material is in full martensite. The stress drops sharply, while the transformation stain remains constant. Then, the material reaches the phase transformation. As in the loading stage, the transformation strain

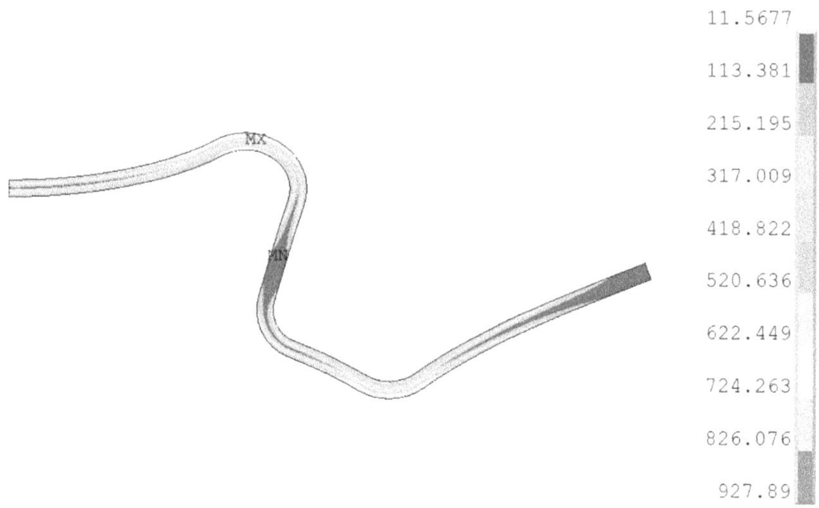

FIGURE 21.14 von-Mises stresses of the orthodontic wire at the end of loading stage (MPa).

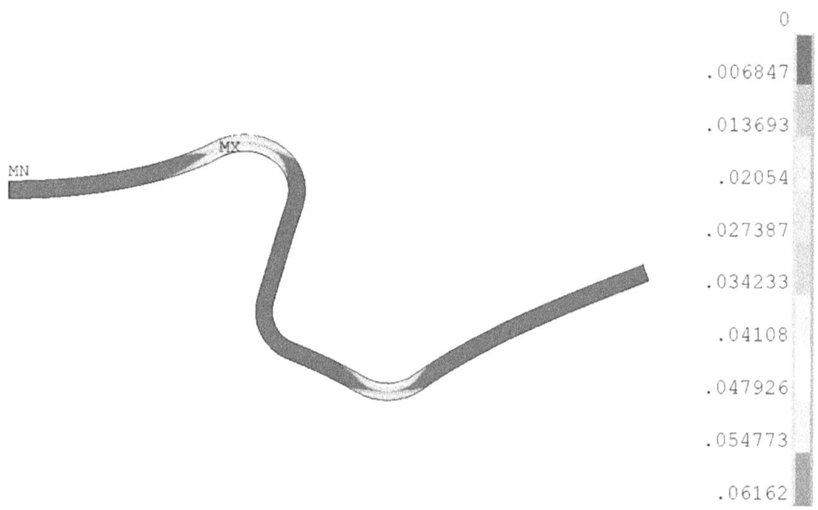

FIGURE 21.15 Transformation strains of the orthodontic wire at the end of loading stage.

drops quickly while the stress reduces slowly. In the final stage of the second stage, the material is full austenite with zero transformation strain.

21.2.5 Discussion

A three-dimensional finite element model of a wire was simulated using the SMA material model for superelasticity. After the wire returns to the starting position, the stress and transformation strain are near zero, which demonstrates the superelasticity of SMA.

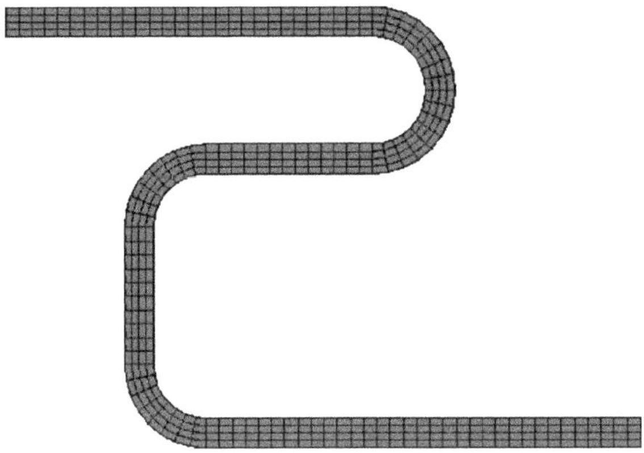

FIGURE 21.16 Deformation at the end of unloading (DMX = 0.2e-3 mm).

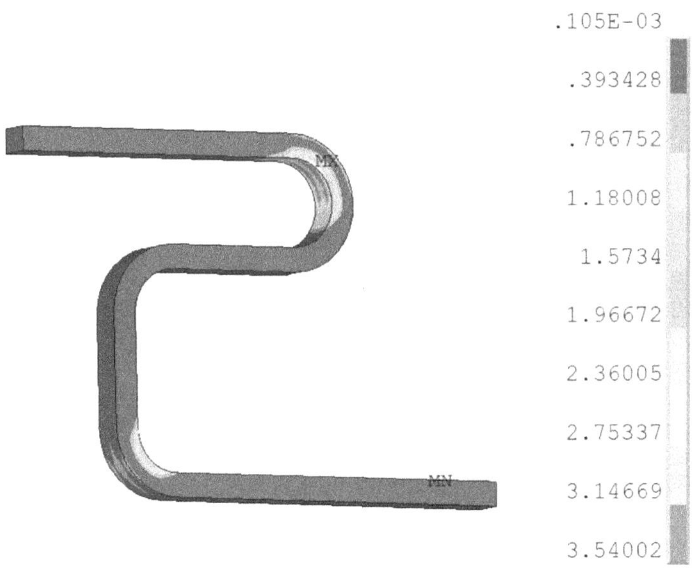

FIGURE 21.17 von-Mises stresses of the wire at the end of unloading (MPa).

However, the stress and transformation strain go through different paths in the first loading and second unloading steps, which are different from those of general elastic materials.

21.2.6 Summary

An orthodontic wire was studied using the SMA material model with superelasticity, which exhibits the superelasticity of SMA.

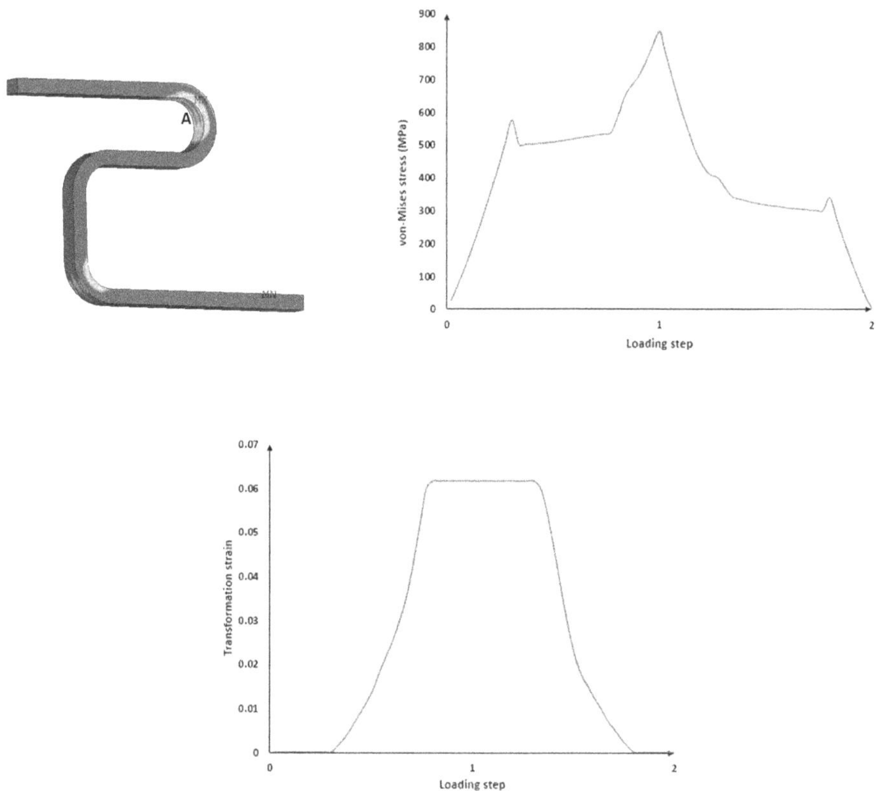

FIGURE 21.18 Time history of stress and transformation strain of element A. (a) Element A, (b) stress of element A vs time, and (c) transformation strain vs time.

21.3 SIMULATION OF A VACUUM-TIGHT SHAPE MEMORY FLANGE

A vacuum-tight shape memory flange was designed for the JET in-vessel inspection system [6], composed of an SMA sleeve and the base connecting the flange to the other apparatus. A silica rod connected a chamber requiring high vacuum (see Figure 21.19). Initially, the diameter of the sleeve is smaller than that of the rod. Under an external force, the sleeve is enlarged to a greater diameter than that of the rod. Thus, the flange was placed around the rod. After heating the flange, the SMA sleeve underwent transformation from martensite to austenite and regained its shape. This caused close contact between the sleeve and the rod, ensuring high vacuum. The whole process was simulated in ANSYS190.

21.3.1 Finite Element Model

Due to the symmetry in two directions, a quarter of the finite element model was built in ANSYS190, including the SMA sleeve, the base, and the rod (see Figure 21.20). The model was meshed using SOLID185 with 2,291 elements and 3,403 nodes.

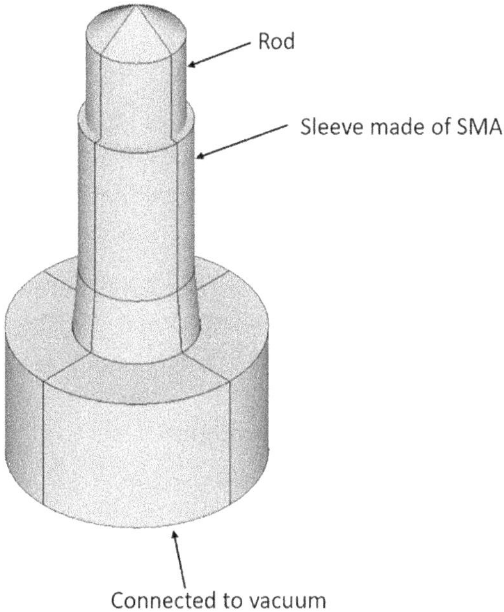

FIGURE 21.19 Illustration of a vacuum-tight shape memory flange.

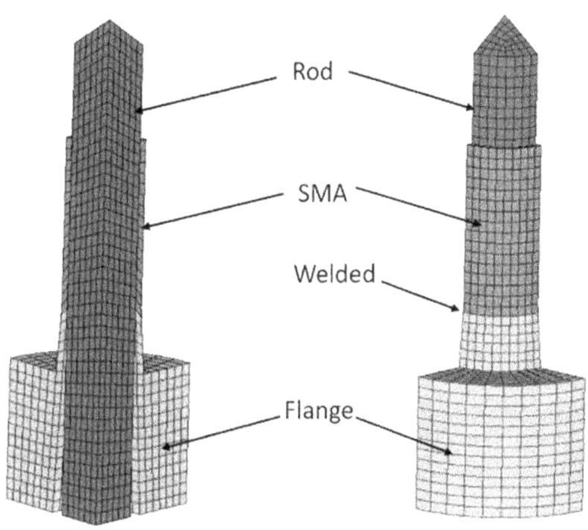

FIGURE 21.20 Finite element model of an SMA flange.

21.3.2 Material Properties

The sleeve was modeled with the SMA material model with the shape memory effect because it involves a temperature change. The material parameters are listed in Table 21.2 [6].

The rod and the base were assumed to be linear elastic with Young's modulus of 74 and 53 MPa, respectively [6].

TABLE 21.2
Material Parameters of Sleeve

E_A (MPa)	v	H	T_0	R	β	$\bar{\epsilon}_L$	E_m (MPa)	m
60,000	0.36	1,000	223	50	2.1	0.04	45,000	0.0

21.3.3 Contact

The sleeve was connected to the base through welding. Thus, they were glued together, sharing common nodes. The contact between the flange and the rod was defined as standard with TARGE170 and CONTA174.

21.3.4 Loadings and Boundary Conditions

The loading went through the following three steps:

1. A pressure of 22 MPa was applied on the sleeve's internal face to enlarge the SMA sleeve while deactivating the contact between the rod and the sleeve. In this stage, the temperature was maintained at 223 K;
2. The rod-sleeve contact was activated, and the applied pressure was removed. The temperature remained at 223 K at this stage; and
3. The temperature was increased to 300 K to induce the phase transformation.

The bottom surface was constrained with all degrees of freedom as boundary conditions.

21.3.5 Solutions

The SMA material model with the shape memory effect was employed. Thus, the full Newton-Raphson with unsymmetrical matrices of elements was defined in the solution setting. In addition, contact existed in the model. To reach convergence, a very small time step of 0.001 was selected for the second and third steps.

21.3.6 Results

Figures 21.21–21.23 plot the displacements and von-Mises stresses for each step. Transformation strains of the second step and third step are presented in Figures 21.24 and 21.25, respectively. At the end of the first step, the sleeve enlarges with a maximum displacement of 5.17 mm. The maximum stress of 362.8 MPa occurs at the connection of the sleeve and the base. In the second step, after the pressure loaded in the first step is removed, the sleeve still has a deformation as high as 4.78 mm, and the maximum stress of 260.7 MPa still exists at the bonding between the sleeve and the base. The remaining deformation of the sleeve due to the existing transformation strain (see Figure 21.24) causes the deformation of the base because of the continuous displacement at the connection between the

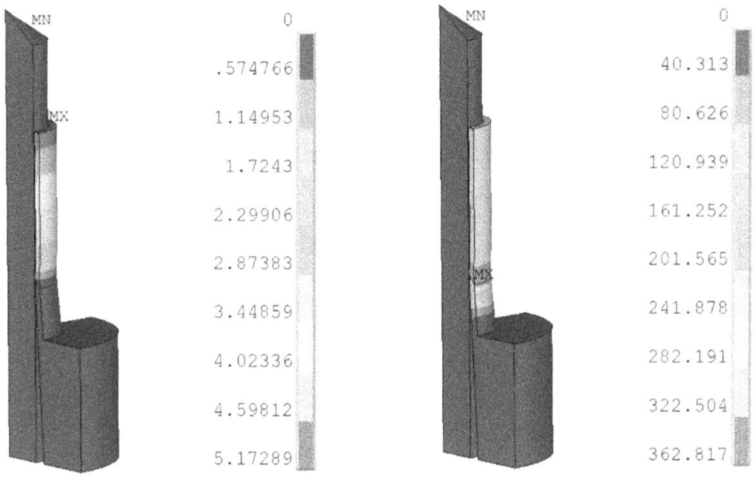

FIGURE 21.21 Results at the end of the first step. (a) Displacement (mm) and (b) von-Mises stress (MPa).

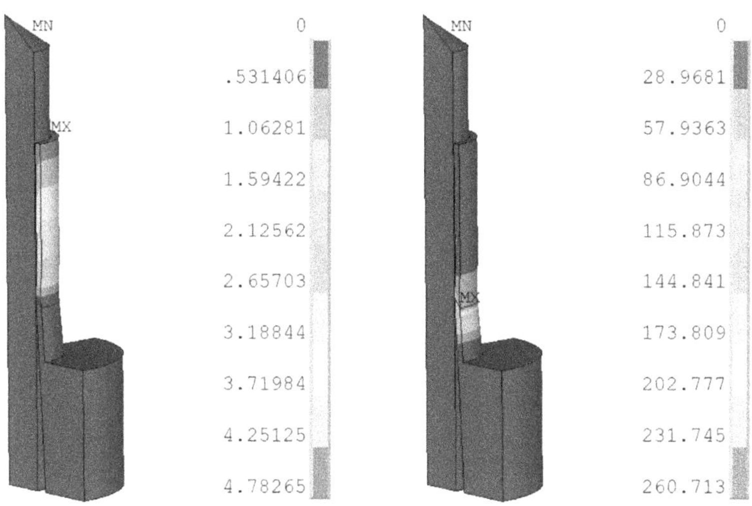

FIGURE 21.22 Results at the end of the second step. (a) Displacement (mm) and (b) von-Mises stress (MPa).

sleeve and the base. That is why a lot of stress occurs at the interface between the sleeve and base. In the third stage, heating induces phase transformation. Thus, the deformation drops significantly to 0.85 mm, and the maximum transformation strain reduces from 0.032 in the second step to 0.0058 in the third step. Therefore, in this step, the stresses in the sleeve are primarily caused by the contact between the rod and the sleeve, with a maximum stress of 145.4 MPa. The contact pressure is illustrated in Figure 21.26.

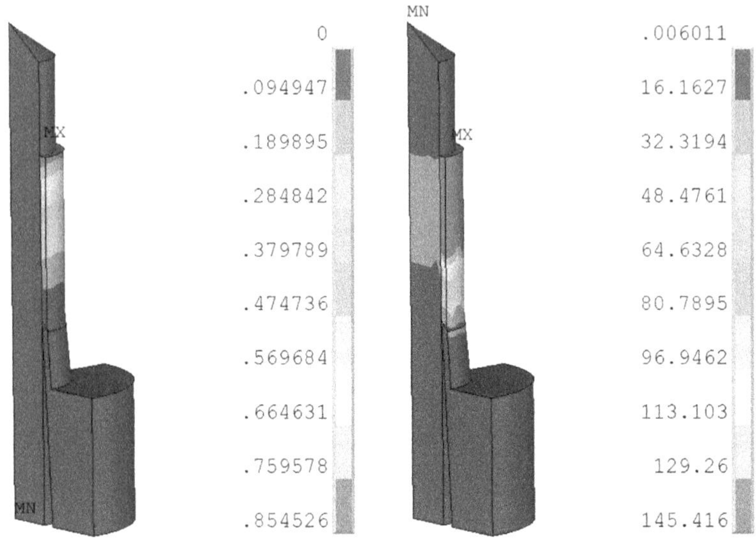

FIGURE 21.23 Results at the end of the third step. (a) Displacement (mm) and (b) von-Mises stress (MPa).

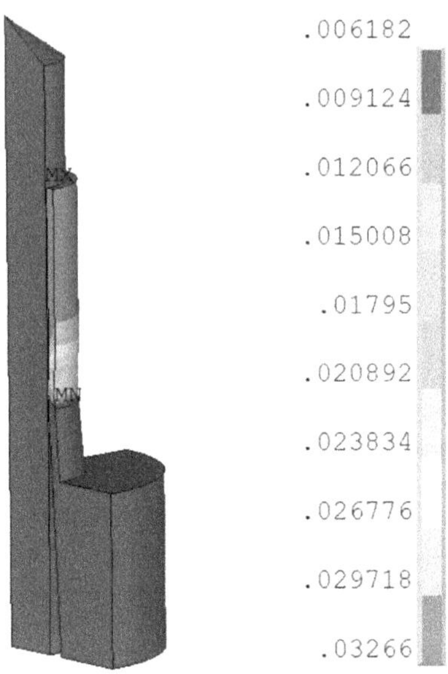

FIGURE 21.24 Transformation strains at the end of step 2.

Figure 21.27 describes the von-Mises stress with time in element B of the base, which indicates that its stress increases to 360 MPa in the first step and gradually drops to 250 MPa in the second step. At the beginning of the third step, it quickly decreases to around 20 MPa. It does not change much at the remainder of the third

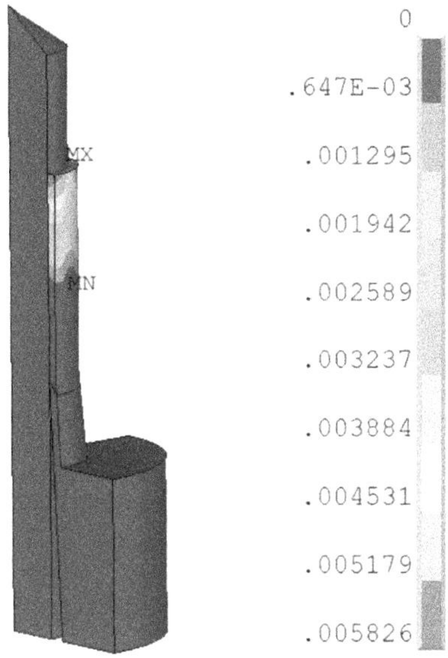

FIGURE 21.25 Transformation strains at the end of step 3.

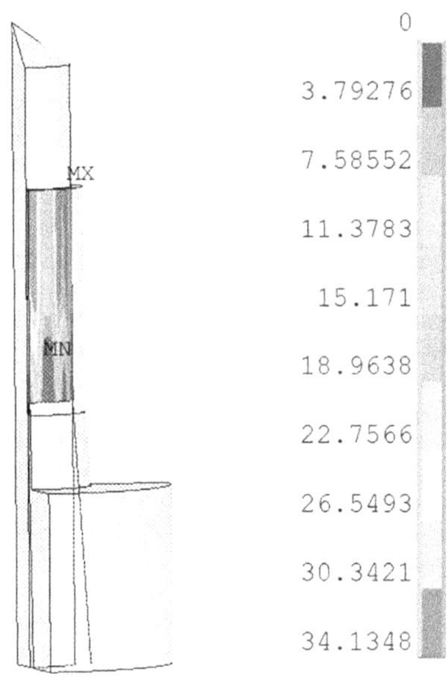

FIGURE 21.26 Contact pressure in the last step (MPa).

step. The variation of the phase transformation strain with stress in element A of the sleeve is plotted in Figure 21.28. It clearly shows that its stress increases in the loading stage in the first step and drops in the unloading stage in the second step. In the third step, with phase transformation, the transformation strain decreases gradually

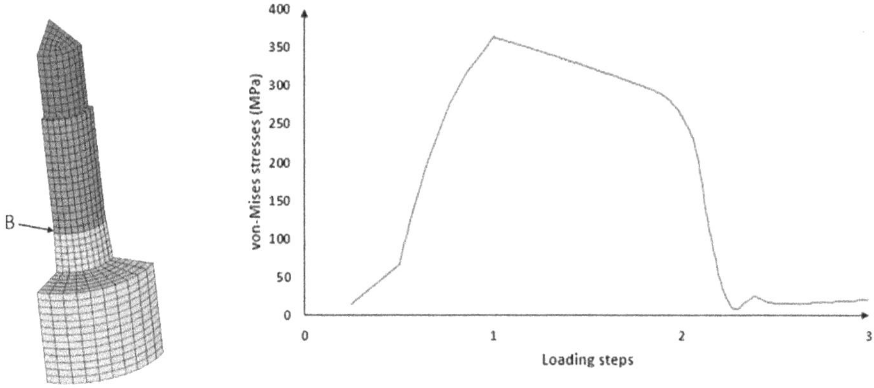

FIGURE 21.27 von-Mises stress of element B with time. (a) Element B and (b) stress of element B at the different loading steps.

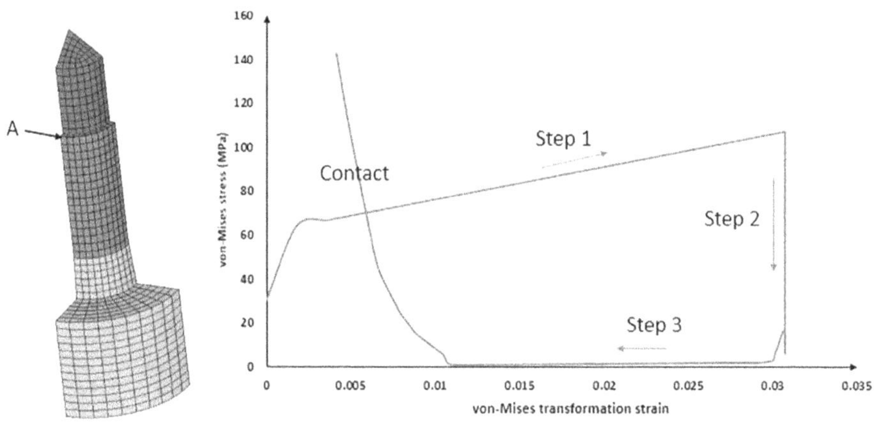

FIGURE 21.28 Phase transformation strain vs stress of element A. (a) Element A and (b) variation of stress of element A with transformation strain.

with near-zero stress. When the sleeve starts to contact the rod, its stress quickly increases.

21.3.7 Discussion

A three-dimensional finite element model of a vacuum-tight flange was built. A part of the flange was modeled using the SMA material model with the shape memory effect. The phase transformation due to heating was simulated, which demonstrated the shape memory effect of SMAs.

This study did not take thermal strain into account because in the model, thermal strain is much less than transformation strain. In addition, phase transformation is strongly nonlinear, and its governing equations are unsymmetrical. Solution convergence requires NROPT, UNSYM, and a very small time step defined in the solution setting.

In this study, the contact between the sleeve and the rod was deactivated and activated using the element birth and death technology, which is widely used in applications such as the modeling of soil excavation.

21.3.8 Summary

A vacuum-tight flange was simulated using the SMA material model with the shape memory effect, which demonstrates the shape memory effect of SMAs.

REFERENCES

[1] Lagoudas, D.C., *Shape Memory Alloys*, Springer, New York, 2008.
[2] ANSYS190 Help Documentation in the help page of product ANSYS190.
[3] Es-Souni, M., Es-Souni, M., and Fischer-Brandies, H., Assessing the biocompatibility of NiTi shape memory alloys used for medical applications, *Analytical and Bioanalytical Chemistry*, Vol.381, 2005, pp. 557–567.
[4] Rahilly, G., and Price, N., Nickel allergy and orthodontics, *Journal of Orthodontics*, Vol. 30, 2003, pp. 171–174.
[5] Auricchio, F., A robust integration-algorithm for a finite-strain shape-memory-alloy superelastic model, *International Journal of Plasticity*, Vol. 17, 2001, pp. 971–990.
[6] Auricchio, F., and Petrini, L., A three-dimensional model describing stress-temperature induced solid phase transformations: solution algorithm and boundary value problems, *International Journal for Numerical Methods in Engineering*, Vol. 61, 2004, pp. 807–836.

22 Simulation of Piezoelectricity

Chapter 22 aims to model piezoelectricity. After introducing piezoelectricity, including the modeling of piezoelectricity in ANSYS, a piezoelectric accelerometer is simulated in ANSYS190 in Section 22.4.

22.1 INTRODUCTION TO PIEZOELECTRICITY

In Greek, "piezo" means pressure. The term piezoelectricity refers to electricity resulting from pressure. Piezoelectricity was studied first by the brothers Jacques Curie (1856–1941) and Pierre Curie (1859–1906) [1,2]. They found that tension and compression in certain crystalline minerals such as tourmaline, quartz, topaz, cane sugar, and Rochelle salt generated voltages of opposite polarity. Later, the Curie brothers discovered the converse effect that one of the voltage-generating crystals exposed to an electric field deformed. After that, people paid more and more attention to the research and application of piezoelectric materials; one of the milestones was the discovery of the phenomenon of ferroelectricity by Valasek. The mixed oxide compound barium titanate, $BaTiO_3$, was the first commercial ferroelectric material that entered the market in 1945.

As one of the branches of crystal physics, piezoelectricity has the following modern applications:

- Frequency control and signal processing, such as mechanical frequency filters, surface acoustic wave devices, and bulk acoustic wave devices;
- Actuators and motors as piezoelectric materials deform when an electric field is applied;
- Sound sensors such as piezoelectric microphones and piezoelectric pickups for electric guitars; and
- Piezoelectric patches used by smart structures to monitor the response of a structure.

22.2 STRUCTURES AND MECHANICAL BEHAVIORS OF PIEZOELECTRIC MATERIALS

Piezoelectric materials are a special type of dielectric in which the electrical and elastic fields are coupled. When a dielectric material is loaded with an external electrical field, electric dipoles are generated due to the interaction of the electrical field with the dielectric structure. The term electric dipole refers to an electro–neutral unit volume. When the center of positive charge within a given region is not in the same position as

DOI: 10.1201/9781003436317-26

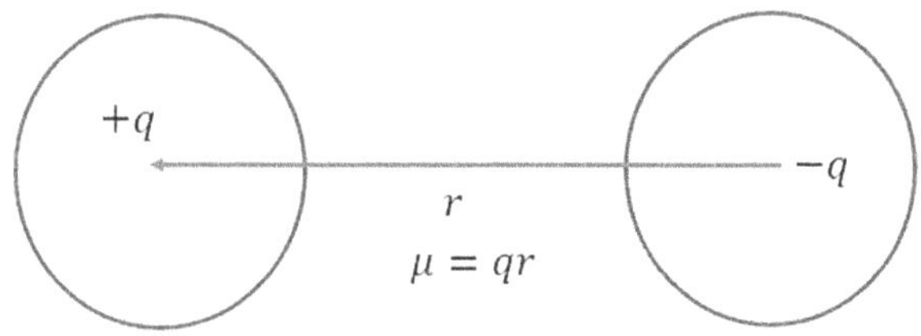

FIGURE 22.1 Dipole moment.

the center of negative charge within the same region, a dipole moment μ generates (see Figure 22.1),

$$\mu = qr \tag{22.1}$$

The polarization of a material is defined as the total dipole moment for a unit volume

$$P = \frac{\Sigma \mu_i}{V} \tag{22.2}$$

where V is the overall volume. Please note that μ_i is a vector sum. When no electric field is loaded, the polar molecules in the material are randomly oriented. Then, the dipole moments cancel out without any net polarization.

When a piezoelectric material is loaded with mechanical stress, the shape of the atomic structure of the crystal deforms. Thus, ions in the structure separate to generate a dipole moment. A net polarization requires that the piezoelectric atomic structure cannot be centrosymmetric because the dipole formed must not be canceled out by other dipoles. When a piezoelectric material is loaded with electric current, the electrical dipoles form to generate the dipole moment and result in deformation.

Ferroelectric materials have spontaneous polarization. Electric dipoles exist in their structure even without an electrical field. A newly-grown single crystal has many domains. Each domain is a homogenous area with individual polarization. All dipole moments in one domain follow the direction of that domain. With many domains along individual polarizations, no overall polarization appears. To have a uniform polarization, a ferroelectric material must go through an additional processing called poling (see Figure 22.2). As the ferroelectric material passes through its Curie temperature, an electric field is loaded on the material to make its spontaneous polarization aligned in one direction. Therefore, the dipole moments in all the domains point in that direction and have a net with the same polarization. After the electric field is removed, most of the dipoles remain nearly aligned. Thus, the material has a remnant polarization.

Overall, some conclusions can be drawn about piezoelectric materials:

1. The coupling of electric and mechanical fields in piezoelectric materials is caused by the specific asymmetric atomic structure of the lattice and the spontaneous polarization at the microstructure level;

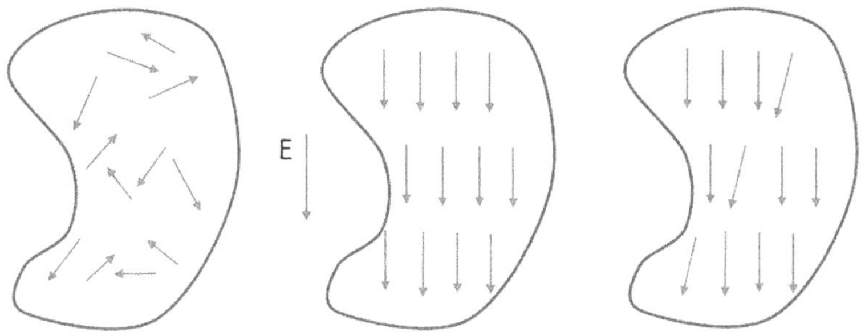

FIGURE 22.2 Structure of ferroelectric materials during poling process [3]. (a) Unpoled ceramic, (b) during poling, and (c) after poling.

2. The combination of the ferroelectric properties of the piezoelectric material with the poling process during manufacture makes these materials widely applicable in modern industry; and
3. The material properties of piezoelectric materials, including elastic stiffness, dielectric constant, and piezoelectric matrix, are anisotropic due to the polarization of the materials.

22.3 CONSTITUTIVE EQUATION OF PIEZOELECTRICITY

The constitutive equation of piezoelectricity is written in the form of the coupling of elasticity with electrostatics as [4]

$$\begin{Bmatrix} \{T\} \\ \{D\} \end{Bmatrix} = \begin{bmatrix} [C] & [e] \\ [e]^T & -[\varepsilon^S] \end{bmatrix} \begin{Bmatrix} \{S\} \\ -\{E\} \end{Bmatrix} \tag{22.3}$$

where T – stress vector

C – elasticity matrix
e – piezoelectric stress matrix
S – elastic strain vector
D – electric flux density vector
ε^S – dielectric matrix
E – electric field intensity vector

The two rows in Equation (22.3) are the constitutive equations for the structural field and electrical field, respectively, except for the coupling terms.

The elasticity matrix $[C]$ has the following form and is defined in ANSYS using TB, ANEL [4]:

$$[C]=\begin{bmatrix} c_{11} & c_{12} & c_{13} & c_{14} & c_{15} & c_{16} \\ & c_{22} & c_{23} & c_{24} & c_{25} & c_{26} \\ & & c_{33} & c_{34} & c_{35} & c_{36} \\ & & & c_{44} & c_{45} & c_{46} \\ & & & & c_{55} & c_{56} \\ & & & & & c_{66} \end{bmatrix} \tag{22.4}$$

The piezoelectric stress matrix [e] is specified as

$$[e]=\begin{bmatrix} e_{11} & e_{12} & e_{13} \\ e_{21} & e_{22} & e_{23} \\ e_{31} & e_{32} & e_{33} \\ e_{41} & e_{42} & e_{43} \\ e_{51} & e_{52} & e_{53} \\ e_{61} & e_{62} & e_{63} \end{bmatrix} \tag{22.5}$$

and input in ANSYS by TB, PIEZ [4].

The orthotropic dielectric matrix $\left[\varepsilon^S\right]$ is written as

$$\left[\varepsilon^S\right]=\begin{bmatrix} \varepsilon_{11} & 0 & 0 \\ 0 & \varepsilon_{22} & 0 \\ 0 & 0 & \varepsilon_{33} \end{bmatrix} \tag{22.6}$$

and described by the MP command in ANSYS [4].

Thus, Equation (22.3) requires an input of the elasticity matrix [C], the piezoelectric stress matrix [e], and the dielectric matrix $\left[\varepsilon^S\right]$ in the finite element models.

22.4 SIMULATION OF PIEZOELECTRIC ACCELEROMETER

One of the applications of piezoelectric materials is the piezoelectric accelerometer, which collects the charge generated by a piezoelectric material while it deforms due to seismic mass vibration. To improve the performance of the piezoelectric accelerometer, the finite element method has been used to study it [5–9]. The goal of this study is to find the relation between the lead zirconate titanate (PZT) film thickness and its performance, which provides a base for the mechanical design of the piezoelectric accelerometer.

22.4.1 Finite Element Model

The finite element model of a piezoelectric thin-film microaccelerometer was built in ANSYS190 (see Figure 22.3), in which a seismic mass in the center was supported by four suspended symmetric beams [8]. Eight piezoelectric transducers were symmetrically

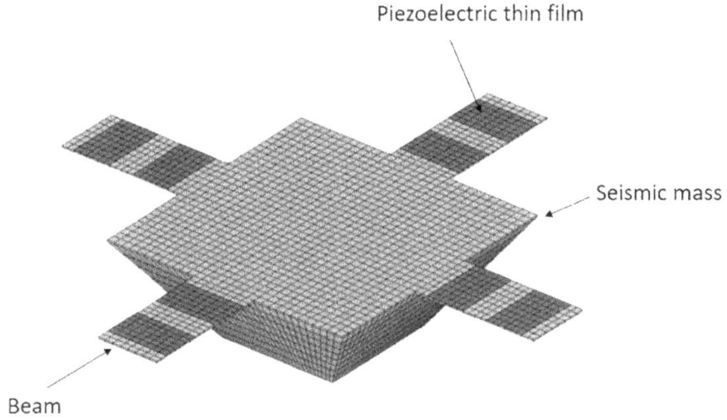

FIGURE 22.3 Finite element model of an accelerometer.

TABLE 22.1
Dimensions of a Piezoelectric Accelerometer

Length of suspension beam	400 μm
Width of suspension beam	200 μm
Thickness of suspension beam	5 μm
Length of seismic mass	800 μm
Thickness of seismic mass	300 μm
Thickness of PZT film	0.5–3 μm

attached to four beams to build the sensing devices. The component dimensions of the accelerometer are listed in Table 22.1. PZT thin films were meshed with SOLID226 with keyoption (1)=1001, while other parts were meshed with SOLID186.

22.4.2 Material Properties

The seismic mass and the beams were assumed to be linear elastic with a Young's modulus of 190 GPa and a Poisson's ratio of 0.3. PZT was defined as anisotropic with a stiffness coefficient matrix as follows [9]:

$$[C]=\begin{bmatrix} 114.25 & 58.294 & 58.525 & 0 & 0 & 0 \\ 58.294 & 114.25 & 58.525 & 0 & 0 & 0 \\ 58.525 & 58.525 & 98.181 & 0 & 0 & 0 \\ 0 & 0 & 0 & 20.747 & 0 & 0 \\ 0 & 0 & 0 & 0 & 20.747 & 0 \\ 0 & 0 & 0 & 0 & 0 & 26.042 \end{bmatrix}(\text{GPa}) \quad (22.7)$$

The corresponding ANSYS commands are as follows:

```
!PIEZOELECTRIC STIFFNESS(ELASTIC) COEFFICIENT MATRIX[C] ( UN/
UM2)  ==> N/M2=1E-6 UN/UM2)
C11=11.425E+4
C12=5.8294E+4
C13=5.8525E+4
C22=11.425E+4
C23=5.8525E+4
C33=9.8181E+4
C44=2.0747E+4
C55=2.0747E+4
C66=2.6042E+4
TB,ANEL,2                              ! INPUT [C] MATRIX FOR PZT
TBDATA,1,C11,C12,C13
TBDATA,7,C22,C23
TBDATA,12,C33
TBDATA,16,C44
TBDATA,19,C55
TBDATA,21,C66
```

The piezoelectric stress constant matrix was given as [8]

$$[e]=\begin{bmatrix} 0 & 0 & 2.3687 \\ 0 & 0 & 7.6006 \\ 0 & 0 & 16.422 \\ 0 & 0 & 0 \\ 10.249 & 0 & 0 \\ 0 & 0 & 0 \end{bmatrix}(\mathrm{PC/um2}) \tag{22.8}$$

It was defined in ANSYS as

```
E13=2.3687
E23=7.6006
E33=16.422
E51=10.249
TB,PIEZ,2                    ! DEFINE PIEZOELECTRIC TABLE FOR
PZT
TBDATA,3,E13                 ! DEFINE MATRIX CONSTANTS
TBDATA,6,E23
TBDATA,9,E33
TBDATA,14,E51
```

The permittivity constant was specified as 10.443e-3, 10.443e-3, and 6.46e-3 pF/um for X, Y, and Z, respectively [8]. The corresponding ANSYS commands were as follows:

```
! DEFINE DIELECTRIC CONSTANT [&] ( PF/UM )  ==> F/M=1E6 PF/UM)
MP,PERX,2,10.443E-3                 ! X DIRECTION PERMITTIVITY
MP,PERY,2,10.443E-3                 ! Y DIRECTION PERMITTIVITY
MP,PERZ,2,6.4605E-3                 ! Z DIRECTION PERMITTIVITY
```

A local coordinate system 12 was defined for the element coordinate of PZT arranged in the Y direction since PZT is anisotropic.

22.4.3 Boundary Conditions and Loadings

Normally, the accelerometer is attached to the body being measured. Thus, the end of each beam was fixed with all degrees of freedom. An acceleration of 1 g was applied to the seismic mass (see Figure 22.4).

22.4.4 Results

Figure 22.5 plots the voltage output after the accelerometer was loaded with 1 g of acceleration. The four beams have a similar distribution where the voltage increases

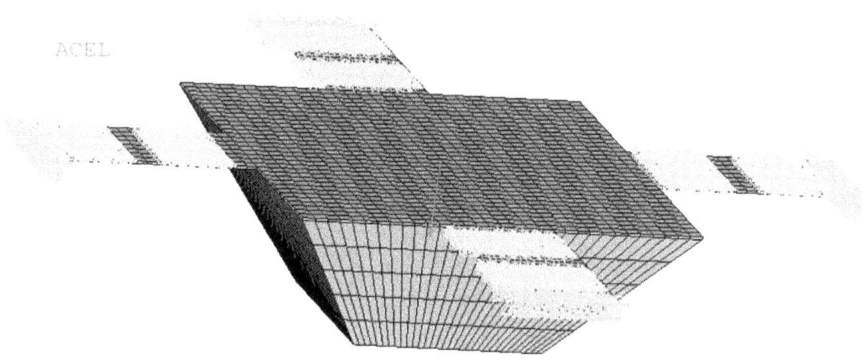

FIGURE 22.4 Boundary conditions and loading.

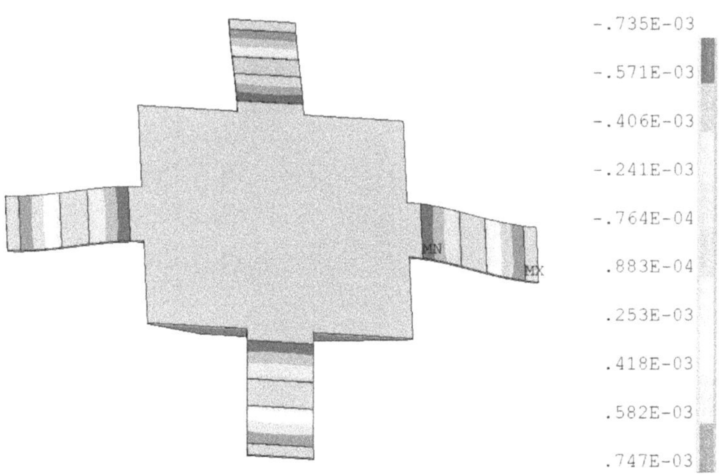

FIGURE 22.5 Voltage output after 1 g acceleration was loaded.

from −0.735e-3V at the inside to 0.747e-3V at the outside. Figure 22.6 illustrates the variation of the PZT film thickness with output voltage; the figure shows that with an increase in the PZT thickness, the voltage output increases to a peak and then drops down.

For comparison, modal analysis was conducted for both the full model (see Figure 22.3) and a quarter of the model under symmetrical conditions (see Figure 22.7). Table 22.2 lists the first five natural frequencies of the full model and one-quarter of the model. Obviously, they are quite different. Figures 22.8–22.12 plot the model shapes of the full model.

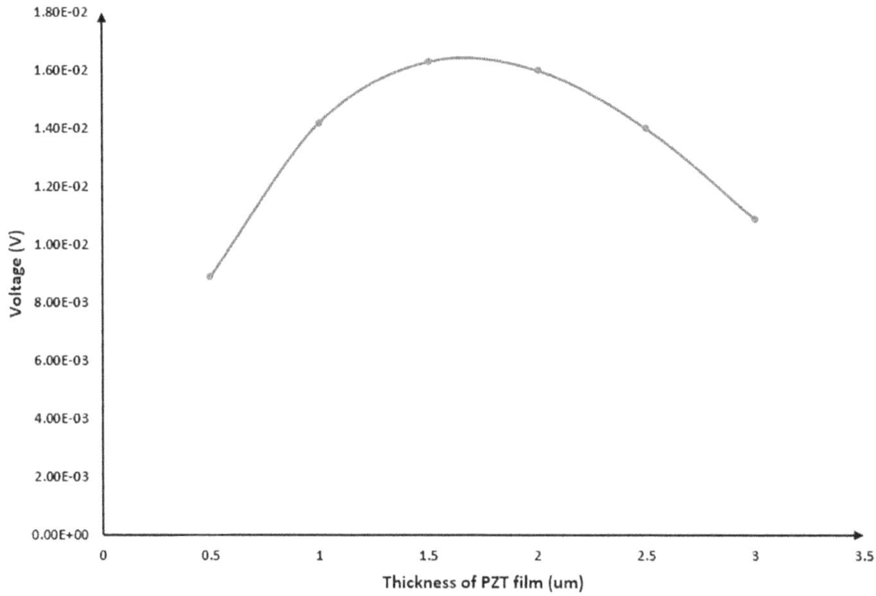

FIGURE 22.6 Variation of voltage with PZT film thickness.

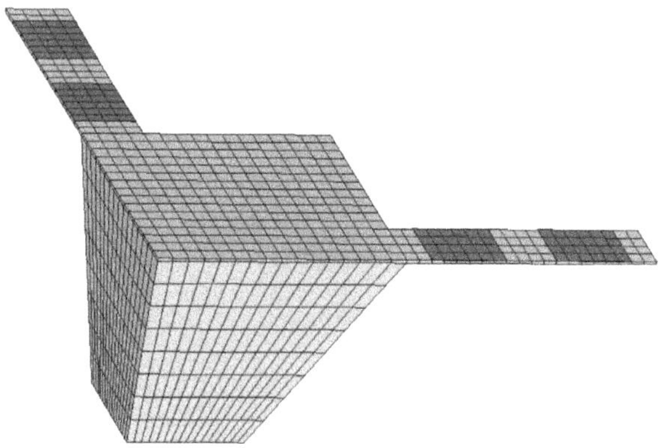

FIGURE 22.7 One-quarter of finite element model of the accelerator.

TABLE 22.2
The First Five Natural Frequencies of One-Quarter of the Model and the Full Model

Natural Frequency (Hz)	One-Quarter of Model	Full Model
First	5594	7108
Second	27326	10624
Third	27369	24743
Fourth	73521	27392
Fifth	73574	27437

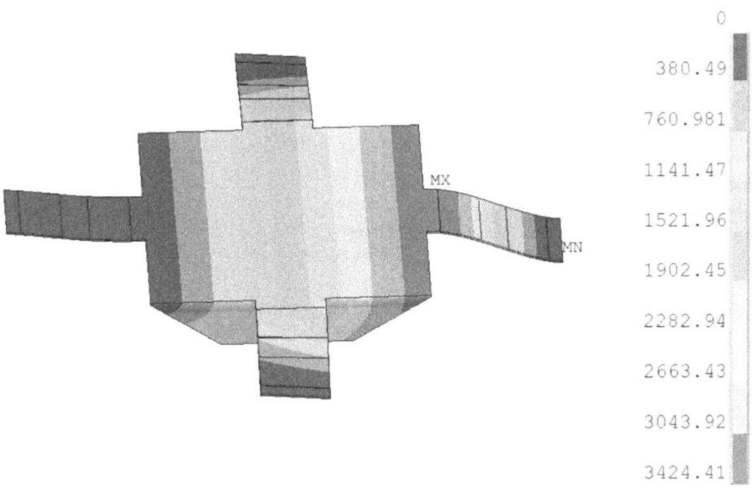

Figure 22.8 The first mode shape of the full model.

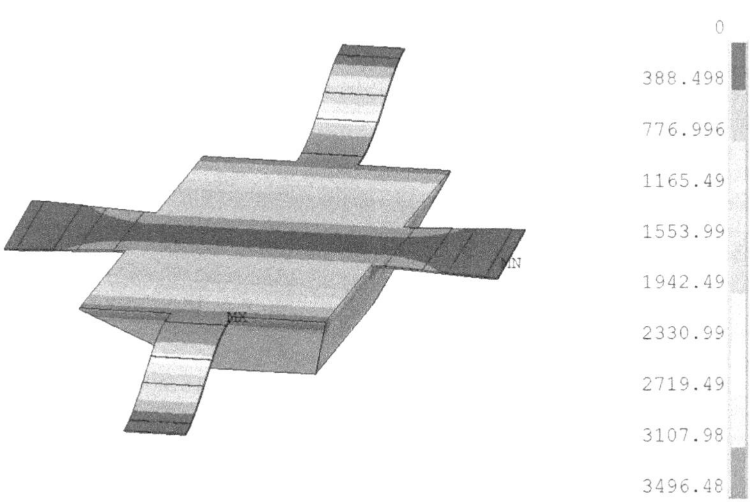

FIGURE 22.9 The second mode shape of the full model.

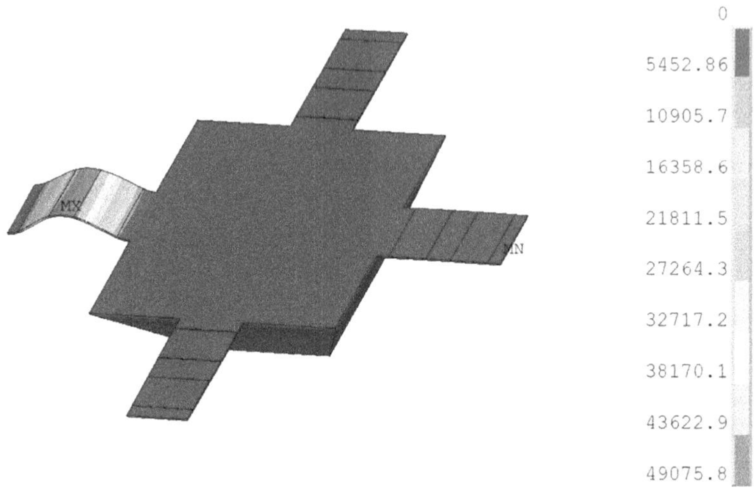

FIGURE 22.10 The third mode shape of the full model.

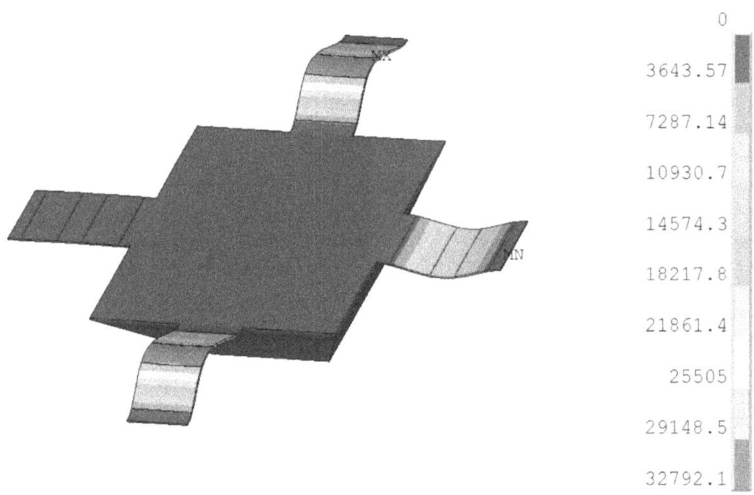

FIGURE 22.11 The fourth mode shape of the full model.

22.4.5 Discussion

Generally, sensitivity and operating frequency range are the two major parameters used to evaluate accelerometer sensors. As sensitivity is defined as voltage output divided by acceleration, in combination with the relation between PZT thickness and voltage output, selecting an appropriate PZT layer thickness can result in high sensitivity of the accelerometer. For the frequency range, compared to the natural frequencies of the full model, one-quarter of the model misses some of the mode shapes

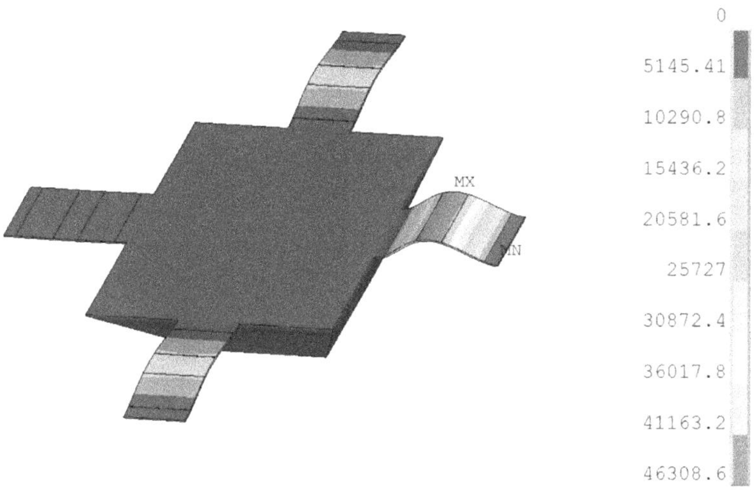

FIGURE 22.12 The fifth mode shape of the full model.

because PZT material is anisotropic, especially since its piezoelectric constants are unsymmetric. Therefore, it would be more appropriate to build a full model, although the structure of PZT is symmetrical.

22.4.6 Summary

A three-dimensional finite element model of a piezoelectric thin-film microaccelerometer was created, and the variation of the PZT layer thickness with the voltage output was studied. The obtained results can be used for guidelines as designing PZT thin-film microaccelerometers. Although the structure of PZT is symmetrical, it would be appropriate to build a full model because the material properties of PZT are unsymmetrical.

REFERENCES

[1] Curie, J., and Curie, P., Développement, par pression, de l'électricité polaire dans les cristaux hémièdres faces ' inclinées, *Comptes Rendus de l'Academie des Sciences, Serie Generale*, Vol. 91, 1880, pp. 294–295.
[2] Curie, J., and Curie, P., Sur l'électricité polaire dans les cristaux hémièdres ' faces inclines, *Comptes Rendus de l'Academie des Sciences, Serie Generale*, Vol. 91, 1880, pp. 383–386.
[3] Dineva, P.S., Gross, D., Muller, R., and Rangelov, T., *Dynamic Fracture of Piezoelectric Materials*, Springer, Cham, 2014.
[4] ANSYS190 Help Documentation in the help page of product ANSYS190.
[5] Gautschi, G., *Piezoelectric Sensorics: Force, Strain, Pressure, Acceleration and Acoustic Emission Sensors, Materials and Amplifiers*, Springer, Berlin, 2002.
[6] Uchino, K., *Piezoelectric Actuators and Ultrasonic Motors*, Kluwer, Boston, 1997.

[7] Macdonald, G.A., A review of low cost accelerometers for vehicle dynamics, *Sensors and Actuators A*, Vol, 21, 1990, pp. 303–307.
[8] Wang, Q-M., Yang, Z., Li, F., and Smolinsk, P., Analysis of thin film piezoelectric micro-accelerometer using analytical and finite element modeling, *Sensors and Actuators, A*, Vol. 113, 2004, pp. 1–11.
[9] Yu, J., and Lan, C., System modeling of microaccelerometer using piezoelectric thin films, *Sensors and Actuators A*, Vol. 88, 2001, pp. 178–186.

23 Nano Materials

Unlike the previously discussed materials, nanomaterials do not follow the traditional mechanics theory. Therefore, most of the material models in ANSYS cannot be applied to nanomaterials. However, the mechanical behaviors of nanomaterials can be studied in ANSYS when the nanomaterials are in the elastic stage. Chapter 23 introduces nanomaterials in Section 23.1 and determines the Young's modulus of Fe particles from the experimental data by using finite element analysis in ANSYS in Section 23.2.

23.1 INTRODUCTION OF NANO

The word "nano" comes from the Latin word *nanus*, meaning dwarf. Therefore, a nanometer is one thousand millionth of a meter (1 nm = 1e-9 m). A nanometer compares to a human hair like a human hair compares to a big house. Nanoscale materials refer to a group of substances with at least one dimension less than approximately 100 nm. Nanomaterials attract more and more interest because of the unique optical, magnetic, electrical, and mechanical properties that exist on that scale. These special properties may have a significant impact in electronics, medicine, and other fields. Some nanomaterials with different shapes, including gold, carbon, metals, meta-oxides, and alloys, are plotted in Figure 23.1.

Based on the modulation dimensionalities of nanomaterials, Siegel classifies nanomaterials as follows: zero-dimensional (atomic clusters), one-dimensional (fibers, wires, and rods), two-dimensional (films, plates, and networks), and three-dimensional (consisting of equiaxed nanometer sized grain) nanostructures.

The physical and chemical properties of the nanomaterials at the nanoscale are much different from those at bulk size, which is due to the nanometer size of the materials resulting in a large fraction of surface atoms, high surface energy, spatial confinement, and reduced imperfections.

23.2 DETERMINATION OF YOUNG'S MODULUS OF FE PARTICLES

23.2.1 Experiment

A compression test was performed on a single-crystal iron particle [1], which is a sphere with an outer diameter of 210 nm composed of Fe and enclosed by a 4-nm shell of $\gamma\text{-}Fe_2O_3$. In the test, the particles were compressed between a silicon substrate and a diamond tip at a constant displacement rate (see Figure 23.2a). During compression, the contact area increased with an increase in the externally applied force until the particles broke suddenly. In order to verify the pure elastic deformation before the collapse, unloading was started prior to the loading reaching the critical point, returned to the starting point, and then reloaded again. The force versus

DOI: 10.1201/9781003436317-27

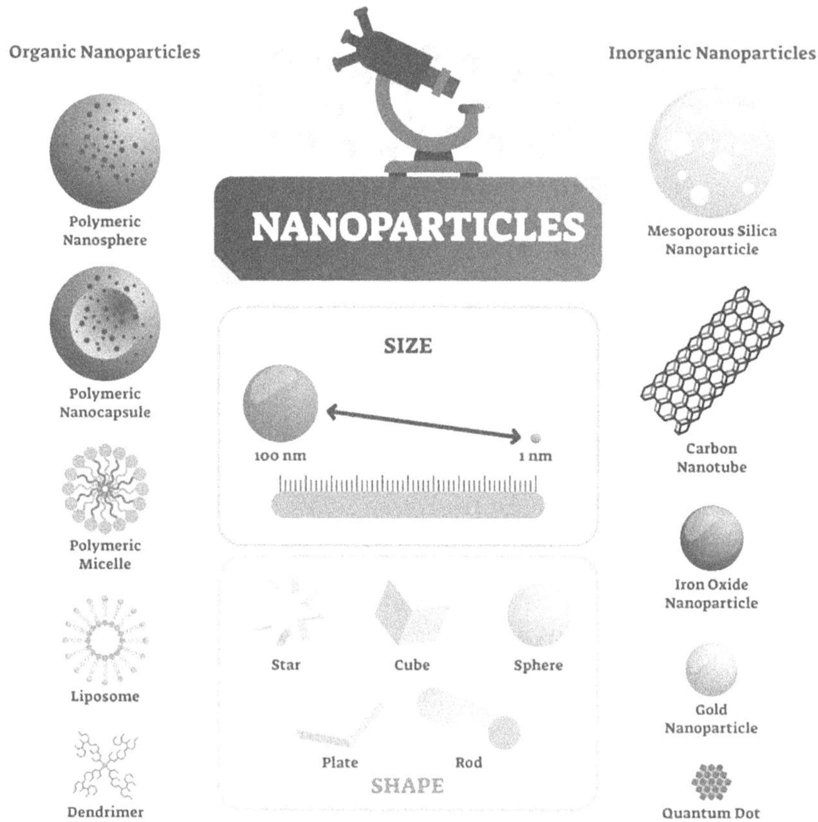

FIGURE 23.1 Nanoparticles. (normalls ©123rf.com.)

the compression displacement repeated in the first and second loadings, which suggests that the particles underwent purely elastic deformation prior to the collapse. Consequently, the Young's modulus of Fe was determined from the experimental reaction forces using the finite element method.

23.2.2 Finite Element Model

The aforementioned experiment was simulated in ANSYS using one-quarter of the finite element model (see Figure 23.2b), in which the sphere was in contact with the silicon substrate and the diamond tip. The silicon substrate and the diamond tip were meshed regularly using SOLID185, while the sphere was meshed freely using SOLID187. The contacts in the model were defined as standard.

23.2.3 Material Properties

The silicon, diamond, and γ-Fe_2O_3 were assumed to be linear elastic with Young's modulus of 170, 1,100, and 220 GPa, respectively [1].

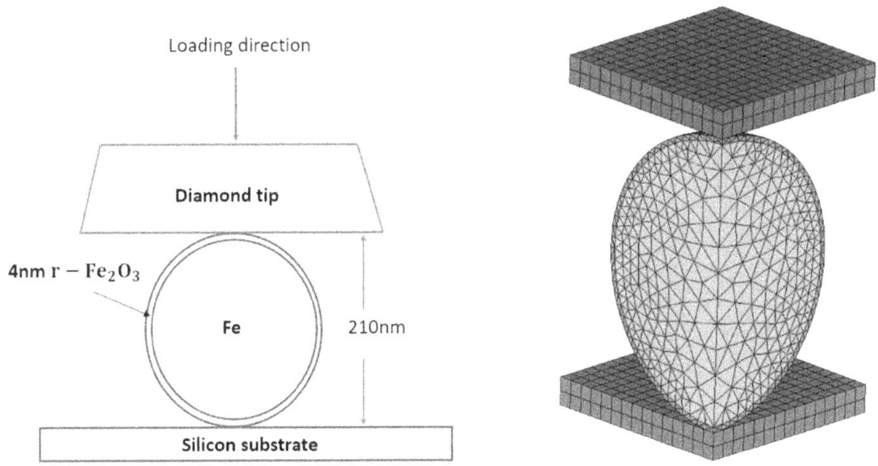

FIGURE 23.2 Experiment and its finite element model. (a) Experiment and (b) a quarter of the model.

23.2.4 Boundary Conditions and Loadings

The bottom of the silicon substrate was fixed with all degrees of freedom, and the top surface of the diamond tip was loaded with the displacement.

23.2.5 Solution

The goal of this study was to find the Young's modulus of the single-crystal iron by using the experimental data for reaction force. As the reaction force versus the displacement is nearly linear except at the starting stage, for simplification, the sum of the reaction forces in the last four substeps was selected for the objective. Initially, 20 GPa was assigned to the single-crystal iron. The whole calculation followed the flow chart shown in Figure 23.3. The experimental loading forces were interpolated from the experimental data.

23.2.6 Results

Computational results indicate that after two iterations, the obtained final reaction force reached the experimental data within 5% error. Then, a Young's modulus of 112.8 GPa defined in the final model was regarded as the material parameter of the single-crystal iron in the experiments. Figure 23.4 plots the computational results against the experimental data, showing that they match well.

23.2.7 Discussion

The scale of the single-crystal iron particles is a nanometer; it is hard to measure the elastic parameters of the particles by using the conventional method. In this study, the Young's modulus of the single-crystal iron particles was obtained from the

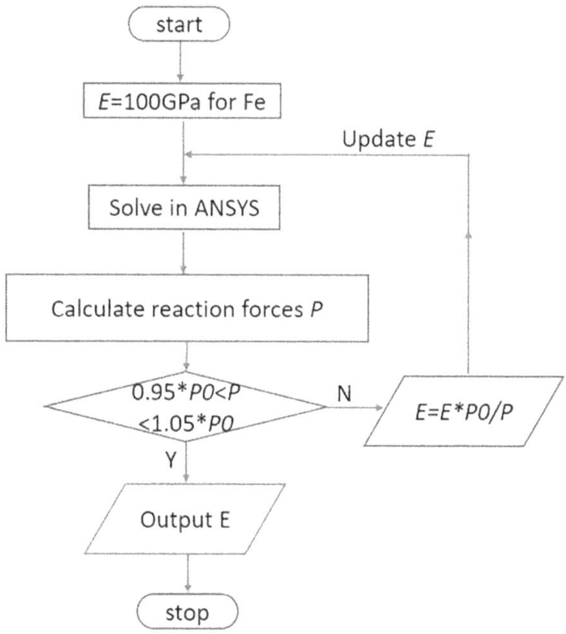

FIGURE 23.3 Flow chart to determine Young's modulus from the experimental data (P0 is the experimental loading force).

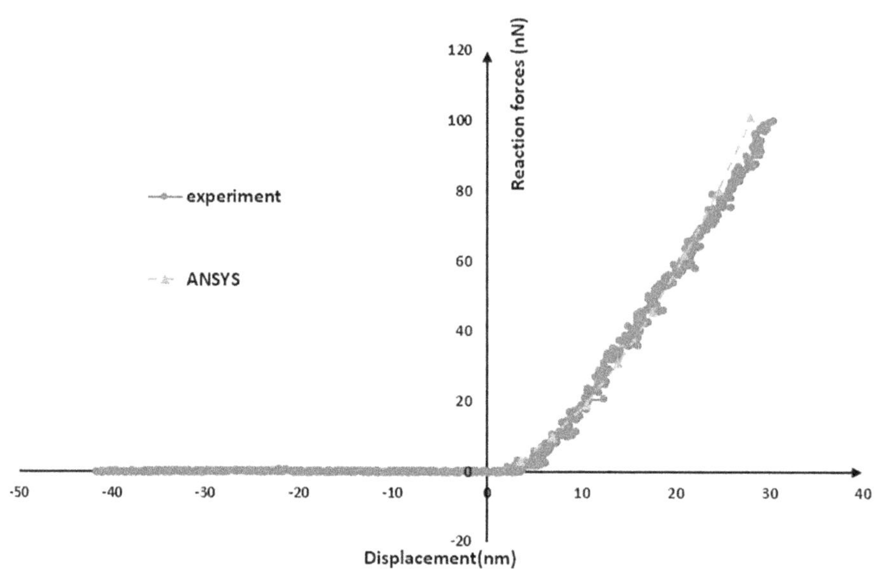

FIGURE 23.4 ANSYS results compared with the experimental data.

contact reaction forces by using the finite element method. The computational results significantly matched the experimental data, validating the proposed method.

This indirect method is always adopted for the nanoscale of mechanical behaviors in many fields. For example, to obtain the cell traction force, the displacement field caused by the cell traction force is first calculated using the image processing technology. Then, the cell traction force is computed inversely from the displacement field by using the finite element method [2].

The difference between the current finite element model and other models in the book is that the material parameter in the current finite element model was updated based on the computational results. Therefore, the optimal process was formed within ANSYS. This method can be expanded for other applications, such as the optimization of mechanical design.

The geometry of the finite element model is on the nanometer scale (1e-9 m). To reduce the computational errors due to this scale, the following units were adopted in this study: mm for length, MPa for pressure, and N for force.

23.2.8 Summary

Young's modulus of single-crystal iron particles was calculated from experimental data using the finite element method. The reaction forces calculated using the defined Young's modulus were very close to the experimental data, which confirms the validity of the obtained Young's modulus.

REFERENCES

[1] Han, W., Huang, L., and Ogata, S., et al., From "smaller is stronger" to "size-independent strength plateau": towards measuring the ideal strength of iron, *Advanced Materials*, Vol. 27, 2015, pp. 3385–3390.

[2] Yang, Z., Lin, J.S., Chen, J., and Wang, J.H.C., Determining substrate displacement and cell traction fields - a new approach, *Journal of Theoretical Biology*, Vol. 242, 2006, pp. 607–616.

Part V

Retrospective

The previous four parts present the modeling of metals, polymers, soils, and modern materials. Based on this content, Part V summarizes material modeling, including association of material properties with the structure, temperature influence, solution setting, definition of material properties, and user subroutine.

DOI: 10.1201/9781003436317-28

24 Retrospective

The first four parts of this book focus on the structures and functions as well as the modeling of some conventional and modern materials. After reviewing these materials, the following conclusions can be drawn:

24.1 CLOSE ASSOCIATION OF MATERIAL PROPERTIES WITH THE STRUCTURE

In a metal, the atoms are bonded together by metallic bonds. The relative sliding of large groups of atoms causes the plastic deformation and ductility of the metal. On the other hand, the macromolecules in a polymer are arranged into a network structure. The weak bonds between macromolecules give the polymer low stiffness and high ductility. When the elongation of macromolecules approaches its limit, the forces within the macromolecules increase dramatically. This explains why polymers have strongly nonlinear behavior under large deformations. Soils are composed of rocks, water, and air. Water and air are in the voids of the soil between the rock particles. When the soils are compacted, the rock particles get closer; because the stiffness of the soil increases with an increase in pore pressure, the stiffness of the soil rises. Consolidation of the soil occurs when water is expelled from the rocks. Different layers are assembled in composites, which makes composites anisotropic. In functionally graded materials, the variation of the structure along the dimensions causes their properties to be functionally graded. Shape memory alloys under loading or temperature change undergo phase transformation, which gives these alloys different mechanical behaviors. Poling makes the ferroelectric material have piezoelectricity. Overall, material properties are strongly related to structure.

24.2 SIGNIFICANT INFLUENCES OF TEMPERATURE AND TIME ON MATERIAL PROPERTIES

Temperature affects the materials in many aspects. First, it produces thermal strain because the material undergoes thermal expansion with a rise in temperature. Also, material properties are a function of the temperature. Young's modulus and plastic parameters of metals always reduce with an increase in temperature. The Young's modulus of polymers is much different between glassy and rubbery regions. Finally, temperature change also affects the phase transformation of shape memory alloys.

If the temperatures of the structure are given, the problem can be solved with the structural elements. However, when the problem is thermal-structure coupled, then it must be solved with the coupled elements, such as CPT elements with keyopt(11)=1 or the coupled 22x elements by the fully coupled method. Section 9.2.2 provides one example of the fully coupled method.

DOI: 10.1201/9781003436317-29

Besides, some material properties are a function of time. Creep exists in all materials such as metals, polymers, and soils. In the metal plastic deformation, a higher rate of strain results in a higher stress with the same strain. Polymers have significant viscoelasticity. Therefore, both time and temperature should be considered in material modeling.

24.3 VARIOUS MATERIALS WITH DIFFERENT SOLUTION CONTROLS

Most plastic material models of metals and soils have a smooth yield surface. Therefore, the Newton-Raphson method is adopted in the nonlinear iteration, and the solution process exhibits very rapid convergence. However, the Mohr-Coulomb and Jointed Rock material models have non-smooth yield surfaces. Therefore, the cutting-plane algorithm using an elastic material tangent is adopted to solve the nonlinear equations for the Mohr-Coulomb and Jointed Rock materials. That requires selecting many substeps in the solution setting of the Mohr-Coulomb and Jointed Rock materials; these substeps show a very slow convergence in the solution process.

24.4 VARIOUS FIELDS WITH DIFFERENT UNITS

The determination of units in a model depends on the application field. For example, a study of such geomechanics problems as tunnel excavation and slope stability analysis requires the selection of the units of force, length, and pressure as N, m, and Pa, respectively. When conducting medical research like the study of the breast tumor, the units of force, length, and pressure (N, mm, and MPa) may be used. Trying to model MEMS and nanomaterials necessitates the adoption of μN, μm and MPa for force, length, and pressure, respectively. Overall, for different scales of problems, the corresponding units for modeling should be selected.

24.5 ANISOTROPIC MATERIAL WITH SYMMETRICAL CONDITIONS

Normally, when the finite element model has symmetrical geometry and loading, it is always simplified to half a model with one direction of symmetry, one-quarter of a model with two directions of symmetry, and one-eighth of a model with three directions of symmetry to speed up the computation. However, in the case of anisotropic materials, such as piezoelectric materials and Jointed Rock materials, the symmetrical condition cannot be applied because the material properties are oriented, not isotropic.

24.6 APPLICATION OF FOUR SOIL MODELS

The Cam Clay model, Drucker-Prager model, Mohr-Coulomb model, and Jointed Rock model are four major soil models. They have different applications for the simulation of soils. For example, the Cam Clay model always works with a porous elastic model to simulate excavation and subsidence of the ground. The Drucker-Prager model and Mohr-Coulomb model can be used to model sand and rock in geotechnics. Mohr-Coulomb better represents the shear failure because the yield function is defined based on the shear planes. However, its yield surface is not smooth. The parameters of the Drucker-Prager model can be chosen in such a way that the Drucker-Prager model is a good and smooth approximation of the Mohr-Coulomb model. The Jointed Rock model is used for modeling anisotropic rocks.

24.7 DEFINITION OF MATERIAL PARAMETERS

Material modeling requires the input of material parameters for various materials. Normally, the material parameters can be found in manuals and references. If not, they can be obtained by curve-fitting experimental data. ANSYS provides curve-fitting functions for the Chaboche model, the creep models, and the hyperelastic models. In addition, the customers can calculate the material parameters from the constitutive law. Some examples are given in Chapters 3, 8, and 21.

24.8 USER SUBROUTINE

ANSYS provides an extensive library of materials. However, when a material cannot be modeled in ANSYS, the customer can develop the user subroutine in ANSYS. Chapters 6 and 12 present two examples, UserMat and UserHyper.

Appendix 1

Input File of Curve-Fitting of the Chaboche Model in Section 3.2

```
/COM, CURVE-FITTING OF THREE-TERM CHABOCHE CURVEFITTING WITH
UNIAXIAL DATA
/COM, UNITS (N, MM, MPA)
*CREATE, LCF.EXP
0.001264272  112.3015873
0.001260138  -34.12698413
0.001164631  -53.57142857
0.001008229  -65.87301587
0.000818357  -78.57142857
0.000491711  -86.9047619
0.000142897  -94.44444444
-0.000285125 -101.984127
-0.00071411  -107.5396825
-0.001140291 -110.7142857
-0.001303876 -112.3015873
-0.001295536 37.6984127
-0.001258384 46.82539683
-0.001190566 53.96825397
-0.001073243 61.11111111
-0.001009193 67.85714286
-0.000813625 76.98412698
-0.000553921 84.52380952
-9.199E-05   95.63492063
0.000283722  100.7936508
0.000643401  106.3492063
0.000980473  109.5238095
0.001153959  111.1111111
*END
/OUT, SCRATCH

/PREP7
TBFT, EADD, 1, UNIA, LCF.EXP
TBFT, FADD, 1, PLAS, CHAB, 3      ! THREE-TERM CHABOCHE MODEL

TBFT, SET, 1, PLAS, CHAB, 3, 1, 1E6
TBFT, SET, 1, PLAS, CHAB, 3, 2, 1E5
TBFT, SET, 1, PLAS, CHAB, 3, 3, 1E4
TBFT, SET, 1, PLAS, CHAB, 3, 4, 1E5
TBFT, SET, 1, PLAS, CHAB, 3, 5, 1E4
```

```
TBFT, SET, 1, PLAS, CHAB, 3, 6, 3000
TBFT, SET, 1, PLAS, CHAB, 3, 7, 55

TBFT, SOLVE, 1, PLAS, CHAB, 3, 0, 1000

/OUT
CFIT, , DFLAG, FITTEDDATLIST
CFIT, , CFNAME, PLCHAB3
CFIT, DEBUG

TBFT, LIST, 1

FINI
```

Appendix 2

Input File of the Ratcheting Model in Section 3.2

```
/COM, SIMULATION OF RATCHETING OF A NOTCHED ROD
/COM, UNITS (N, MM, MPA)
/PREP7
RECTNG, 0, 100, 0, 200
CYL4, 100, 0, 50
ASBA, 1, 2
ET, 1, 182
KEYOPT, 1, 3, 1
MSHAPE, 0, 2D
LESIZE, 3, , , 8, , , , , 1
LESIZE, 4, , , 10, , , , , 1
LESIZE, 10, , , 8, , , , , 1
LESIZE, 6, , , 10, , , , , 1
AMESH, 3

EP=5.47E4
NU=0.3
MP, EX, 1, EP
MP, NUXY, 1, NU
TB, NLISO, 1, , 2, 5       ! NONLINEAR ISOTROPIC HARDENING
POWER_N=0.1
SIGMA_Y=67.5
TBDATA, 1, SIGMA_Y, POWER_N
TB, CHAB, 1, , 3    ! CHABOCHE MODEL
TBDATA, 1, 67.5, 1E6, 9.37E4, 1E4, 1E5,
TBDATA, 6, 4.1E4, 1.1E3

NSEL, S, LOC, X, 0
D, ALL, UX
NSEL, S, LOC, Y, 0
D, ALL, UY
ALLSEL
FINISH

/SOLU
NSUBST, 50, 1000, 20
OUTRES, ALL, ALL
*DO, I, 1, 10 ! CYCLIC LOADING
TIME, 5+(I-1)*8
```

```
NSEL, S, LOC, Y, 200
SF, ALL, PRES, -50
ALLSEL
SOLV
TIME, 8+(I-1)*8
NSEL, S, LOC, Y, 200
SF, ALL, PRES, 30
ALLSEL
SOLV
*ENDDO
FINISH
/POST26
ESOL, 2, 75, 4, S, Y,
ESOL, 3, 75, 4, EPEL, Y,
ESOL, 4, 75, 4, EPPL, Y,
ADD, 5, 3, 4
XVAR, 5
PLVAR, 2, , , , , , , , ,
```

Appendix 3

Input File of the Forming Process Model in Section 4.1

```
/COM, SIMULATION OF FORMING PROCESS OF A METAL SHEET
/COM, UNITS (N, MM, MPA)
/PREP7
!***********MATERIAL PROPERTIES*************
MP, EX, 1, 149650
MP, NUXY, 1, 0.33
TB, CHAB, 1, , 1 ! DEFINE CHABOCHE MATERIAL DATA
TBDATA, 1, 1.53E2, 62511, 1.1/311*62511
TB, RATE, 1, , 6, EVH ! DEFINE RATE DEPENDENT MATERIAL DATA
TBDATA, 1, 1.53E2, 0, -1.53E2, 317, 1/7.7, 1150

K, 1, 49, 50
K, 2, 49, 6
K, 3, 44, 1
K, 4, 0, 1
K, 5, 44, 6
L, 2, 1
L, 4, 3
LARC, 3, 2, 5, 5,

K, 6, 100, 0
K, 7, 55, 0
K, 8, 50, -5
K, 9, 55, -5
LARC, 7, 8, 9, 5
K, 10, 50, -25
K, 11, 45, -30
K, 12, 45, -25
LARC, 10, 11, 12, 5
K, 13, 0, -30
L, 11, 13
L, 8, 10
L, 6, 7

RECTNG, 0, 95, 0, 1
ET, 1, 182
KEYOPT, 1, 3, 1
KEYOPT, 1, 6, 1
ESIZE, 0.5
AMESH, 1
```

```
ET, 2, 169
ET, 3, 172
R, 2, , , ,

TYPE, 2
REAL, 2
ESIZE, 0.25
LMESH, 1, 3
ALLSEL, ALL

TYPE, 3
REAL, 2
LSEL, S, LINE, , 11
NSLL, S, 1
ESLN, S
ESURF
ALLSEL

ET, 4, 169
ET, 5, 172
R, 3, , , ,

TYPE, 4
REAL, 3
ESIZE, 0.5
LMESH, 4, 8
ALLSEL, ALL

TYPE, 5
REAL, 3
LSEL, S, LINE, , 9
NSLL, S, 1
NSEL, R, LOC, X, 0, 95
ESLN, S
ESURF
ALLSEL

K, 21, 55, 1
K, 22, 100, 1
L, 21, 22

ET, 6, 169
ET, 7, 172
R, 4, , , ,

TYPE, 6
REAL, 4
ESIZE, 0.5
LMESH, 13
ALLSEL, ALL
```

```
TYPE, 7
REAL, 4
LSEL, S, LINE, , 11
NSLL, S, 1
ESLN, S
ESURF
ALLSEL

ESEL, S, TYPE, , 1
NSLE, S
NSEL, R, LOC, X, 90, 95
D, ALL, UY
ESEL, S, TYPE, , 2
NSLE, S
D, ALL, UX, 0
D, ALL, UY, -30
ALLSEL
ESEL, S, TYPE, , 4, 6, 2
NSLE, S
D, ALL, ALL
ALLSEL

!--------------CREATE COMPONENTS--------------
ESEL, S, ENAME, , 182
CM, COMP1, ELEM ! CREATE NLAD COMPONENT
ALLS
FINISH

/SOLU
!-----------GENERAL SOLUTION SETTINGS---------
NLGEOM, ON
RATE, ON
ERESX, NO
OUTRES, ALL, ALL
!---------------NLAD SETTINGS----------------- ! CHECK NLAD
PARAMETERS SECTION ABOVE FOR VALUES
NLAD, COMP1, ADD, MESH, SKEW, 0.95 ! NLAD WITH MESH-QUALITY-
BASED CRITERION
NLAD, COMP1, ON, , , 1, 0, 1

!---------------STEP SETTINGS-----------------
TIME, 1
NSUBST, 300, 10000, 100
NEQIT, 120
ALLS
SOLVE
FINI

/POST1
*GET, NSTP, ACTIVE, 0, SOLU, NCMSS
NSTP=103
```

```
*DIM, RFTOL, ARRAY, NSTP, 2
*DO, I, 1, NSTP
ALLSEL
SET, 1, I
*GET, RF1, NODE, 574, RF, FY

RFTOL(I, 2)=RF1
*GET, RFTOL(I, 1), ACTIVE, 0, SET, TIME
*ENDDO
*STATUS, RFTOL
```

Appendix 4
Input File of the Bolt Model under Pretension in Section 4.2

```
/COM, SIMULATION OF A BOLT UNDER PRETENSION (CREEP)
/COM, UNITS (LBF, INCH, PSI)
/PREP7
ET, 1, 187
MP, EX, 1, 14.665E6
MP, PRXY, 1, 0.30
TB, PLAS, 1, 1, 2, MISO    ! MULTILINEAR ISOTROPIC HARDENING
TBPT, DEFI, 0, 29.33E3
TBPT, DEFI, 1.59E-3, 50E3

MP, EX, 2, 42E6
MP, PRXY, 2, 0.30
TB, PLAS, 2, 1, 2, MISO    ! MULTILINEAR ISOTROPIC HARDENING
TBPT, DEFI, 0, 29.33E3*3
TBPT, DEFI, 1.59E-3, 50E3*3
TB, CREEP, 2, 1, 4, 2
TBDATA, 1, 0.54E-16, 2.3, -0.6

CYLIND, 0.5, , -0.25, 0, 0, 180
CYLIND, 0.5, , 1, 1.25, 0, 180
CYLIND, 0.25, , 0, 1, 0, 180
WPOFF, .05
CYLIND, 0.35, 1, 0, 0.75, 0, 180
WPOFF, -.1
CYLIND, 0.35, 1, 0.75, 1, 0, 180
WPSTYLE, , , , , , , , 0
VGLUE, ALL
NUMC, ALL
MAT, 1
SMRT, OFF
VMESH, 4, 5
MAT, 2
VMESH, 1, 3
PSMESH, , BOLT, , VOLU, 1, 0, Z, 0.5, , , , ELEMS
FINISH

/SOLU
TIME, 1E-4
KBC, 1
```

```
RATE, 0
NSEL, S, LOC, Y, 0
D, ALL, UY, 0
ALLSEL
D, NODE(0, 0, -0.25), UX, 0
D, NODE(0, 0, -0.25), UZ, 0
D, NODE(0, 0, 1.25), UX, 0
RESCONTROL, LINEAR, ALL, 1
SLOAD, 1, PL01, TINY, DISP, 0.01, 1, 2
SOLVE
TIME, 1.5E6
RATE, 1
NSUBST, 100
OUTRES, ALL, ALL
SOLVE
FINISH
```

Appendix 5

Input File of the Combustion Chamber Model in Section 5.2

```
/COM, SIMULATION OF A COMBUSTION CHAMBER
/COM, UNITS (N, MM, MPA)
/PREP7
K, 1, -320, -535
K, 2, 0, -535
K, 3, 0, 0
K, 4, -320, 0

A, 1, 2, 3, 4

CYL4, -110, -265, 100
ASBA, 1, 2
CYL4, 0, 0, 80
ASBA, 3, 1
ET, 1, 182
KEYOPT, 1, 3, 2
KEYOPT, 1, 1, 1

ESIZE, 25
AMESH, ALL

ALLSEL
FLST, 5, 13, 2, ORDE, 8
FITEM, 5, 14
FITEM, 5, -18
FITEM, 5, 137
FITEM, 5, -138
FITEM, 5, 147
FITEM, 5, -148
FITEM, 5, 159
FITEM, 5, -162
ESEL, S, , , P51X
FLST, 5, 13, 2, ORDE, 8
FITEM, 5, 14
FITEM, 5, -18
FITEM, 5, 137
FITEM, 5, -138
FITEM, 5, 147
FITEM, 5, -148
```

```
FITEM, 5, 159
FITEM, 5, -162
CM, _Y, ELEM
ESEL, , , , P51X
CM, _Y1, ELEM
CMSEL, S, _Y
CMDELE, _Y
EREF, _Y1, , , 1, 0, 1, 1
CMDELE, _Y1
ALLSEL
EPLOT

TB, ELAS, 1, 3 !    ELASTIC PROPERTIES
TBTEMP, 20
TBDATA, 1, 76000, 0.33
TBTEMP, 100
TBDATA, 1, 72000, 0.33
TBTEMP, 300
TBDATA, 1, 60000, 0.33

TB, CHAB, 1, 4, 1   ! CHABOCHE MODEL
TBTEMP, 20
TBDATA, 1, 80, 21700, 155
TBTEMP, 100
TBDATA, 1, 63, 22000, 200
TBTEMP, 200
TBDATA, 1, 52, 8460, 310
TBTEMP, 300
TBDATA, 1, 4, 3780, 420

TB, RATE, 1, 3, , PERZYNA  ! RATE TABLE
TBTEMP, 20
TBDATA, 1, 1/8.5, 1E-4
TBTEMP, 200
TBDATA, 1, 1/8.5, 3E-6
TBTEMP, 300
TBDATA, 1, 1/8.5, 1E-10

MP, ALPX, 1, 6E-6

TB, PLAS, 1, 4, , MISO    ! MULTILINEAR ISOTROPIC HARDENING
TBTEMP, 20
TBPT, DEFI, 0, 80
TBPT, DEFI, 0.1, 200
TBTEMP, 100
TBPT, DEFI, 0, 63
TBPT, DEFI, 0.1, 157
TBTEMP, 200
TBPT, DEFI, 0, 52
TBPT, DEFI, 0.1, 130
TBTEMP, 300
```

```
TBPT, DEFI, 0, 4
TBPT, DEFI, 0.1, 10

NSEL, S, LOC, X, 0
NSEL, A, LOC, X, -320
D, ALL, UX, 0
NSEL, S, LOC, Y, 0
NSEL, A, LOC, Y, -535
D, ALL, UY, 0
ALLSEL

TREF, 0
FINISH

/SOLU
NSUBST, 100, 1000, 30
KBC, 0
RATE, ON
OUTRES, ALL, ALL
TIME, 80
BF, ALL, TEMP, 280
SOLVE

TIME, 160
BF, ALL, TEMP, 80
SOLVE

TIME, 240
BF, ALL, TEMP, 280
SOLVE

TIME, 320
BF, ALL, TEMP, 80
SOLVE

TIME, 400
BF, ALL, TEMP, 280
SOLVE

TIME, 500
BF, ALL, TEMP, 80
SOLVE

FINISH
```

Appendix 6

UserMat of Strain-Hardening Model in Section 6.2

```
      subroutine usermat(
& matId, elemId,kDomIntPt, kLayer, kSectPt,
& ldstep,isubst,keycut,
& nDirect,nShear,ncomp,nStatev,nProp,
& Time,dTime,Temp,dTemp,
& stress,ustatev,dsdePl,sedEl,sedPl,epseq,
& Strain,dStrain, epsPl, prop, coords,
& rotateM, defGrad_t, defGrad,
& tsstif, epsZZ,
& var1, var2, var3, var4, var5,
& var6, var7, var8)

c*****************************************************************
#include "impcom.inc"
EXTERNAL
& CSS,
& Rvoight1,
& OUTdsdep0,
& Tnorm,
& TnormE

c
      INTEGER
&     matId, elemId,
&     kDomIntPt, kLayer, kSectPt,
&     ldstep,isubst,keycut,
& nDirect,nShear,ncomp,nStatev,nProp
      DOUBLE PRECISION
&     Time, dTime, Temp, dTemp,
&     sedEl, sedPl, epseq, epsZZ
      DOUBLE PRECISION
&     stress (ncomp ), ustatev (nStatev),
&     dsdePl (ncomp,ncomp),
&     Strain (ncomp ), dStrain (ncomp ),
&     epsPl (ncomp ), prop (nProp ),
&     coords (3), rotateM (3,3),
&     defGrad (3,3), defGrad_t(3,3),
&     tsstif (2)
c
c*************** User defined part ******************************
```

```
DOUBLE PRECISION EE, vv, Ks, ns, Strain1(6), tinny, tsn
DOUBLE PRECISION equ_de, equ_e, tss, dsde, sn(6), EY,Tnorm
DOUBLE PRECISION E(3,3), dE(3,3), ds(3,3), tLam, tmuA, tsn1,
delE
DOUBLE PRECISION var1, var2, var3, var4, var5, var6, var7,
var8
DOUBLE PRECISION tds, ds6(6), TnormE
INTEGER i, np
c convergernce criteria
tinny=1e-10
c get material properties
EE=prop(1)
vv=prop(2)
Ks=prop(3)
ns=prop(4)

c input stress at time t
sn(1) = ustatev(1)
sn(2) = ustatev(2)
sn(3) = ustatev(3)
sn(4) = ustatev(4)
sn(5) = ustatev(5)
sn(6) = ustatev(6)
tsn = Tnorm(sn)

c strain input
DO i=1,6
Strain1(i) = Strain(i) + dStrain(i)
ENDDO

c convert 3D into 1D problem
CALL Rvoight1(Strain1,E)
CALL Rvoight1(dStrain,dE)

equ_de=TnormE(dStrain)/(1.0+vv)
equ_e=TnormE(Strain1)/(1.0+vv)

IF(tsn.LT.tinny)Then
tss=0
else
tss=tsn
endif
c solve 1D problem using Newton-Raphson method
np=0
dsde = EE/(1.0 + ns*Ks*EE*(tss)**(ns-1))
tsn1 = tsn + dsde*equ_de
10     delE=equ_e-tsn1/EE-Ks*(tsn1)**ns
if(np.gt.25) then
keycut=1
goto 20
elseif(abs(delE).GT.(1e-5*equ_e))Then
```

```
tss=tsn1
dsde = EE/(1.0 + ns*Ks*EE*(tss)**(ns-1))
tsn1 = tsn1 + dsde*delE
np=np+1
goto 10
else
c find the equivalent Young's modulus EY
EY = dsde
endif
tLam=EY*vv/(1-2*vv)/(1+vv)
tmuA=EY/2/(1+vv)
c find delS at dE
CALL CSS(tLam,tmuA,dE,ds)
ds6(1)=ds(1,1)
ds6(2)=ds(2,2)
ds6(3)=ds(3,3)
ds6(4)=ds(1,2)
ds6(5)=ds(2,3)
ds6(6)=ds(1,3)
tds=Tnorm(ds6)
CALL OUTdsdep0(dsdePl,EY,vv)
c sn1=sn + delS(after scaling)
if(tds.GT.tinny)Then
stress(1)=ds(1,1)*abs(tsn1-tsn)/tds+sn(1)
stress(2)=ds(2,2)*abs(tsn1-tsn)/tds+sn(2)
stress(3)=ds(3,3)*abs(tsn1-tsn)/tds+sn(3)
stress(4)=ds(1,2)*abs(tsn1-tsn)/tds+sn(4)
stress(5)=ds(2,3)*abs(tsn1-tsn)/tds+sn(5)
stress(6)=ds(1,3)*abs(tsn1-tsn)/tds+sn(6)
else
stress(1)=ds(1,1)+sn(1)
stress(2)=ds(2,2)+sn(2)
stress(3)=ds(3,3)+sn(3)
stress(4)=ds(1,2)+sn(4)
stress(5)=ds(2,3)+sn(5)
stress(6)=ds(1,3)+sn(6)
endif

c save stress in the state variables
ustatev(1)=stress(1)
ustatev(2)=stress(2)
ustatev(3)=stress(3)
ustatev(4)=stress(4)
ustatev(5)=stress(5)
ustatev(6)=stress(6)
20 continue
Return
END

FUNCTION Tnorm(Tens)
DOUBLE PRECISION Tens(6),Tnorm
```

```
Tnorm=0.0d0
Tnorm=Tnorm + (Tens(1)-Tens(2))**2
Tnorm=Tnorm + (Tens(2)-Tens(3))**2
Tnorm=Tnorm + (Tens(1)-Tens(3))**2
Tnorm=Tnorm + 6*(Tens(4)**2+Tens(5)**2+Tens(6)**2)

Tnorm=sqrt(Tnorm/2.0)
RETURN
END

FUNCTION TnormE(Tens)
DOUBLE PRECISION Tens(6),Tnorm, TnormE
Tnorm=0.0d0
Tnorm=Tnorm + (Tens(1)-Tens(2))**2
Tnorm=Tnorm + (Tens(2)-Tens(3))**2
Tnorm=Tnorm + (Tens(1)-Tens(3))**2
Tnorm=Tnorm + 1.5*(Tens(4)**2+Tens(5)**2+Tens(6)**2)
TnormE=sqrt(Tnorm/2.0)
RETURN
END

SUBROUTINE OUTdsdep0(dsde,EY,vv)
DOUBLE PRECISION dsde(6,6),EY,vv
INTEGER i,j
DO i=1,6
DO j=1,6
dsde(i,j)=0.0d0
ENDDO
ENDDO
dsde(1,1)=EY/(1+vv)/(1-2*vv)*(1-vv)
dsde(1,2)=EY/(1+vv)/(1-2*vv)*vv
dsde(1,3)=EY/(1+vv)/(1-2*vv)*vv
dsde(2,1)=EY/(1+vv)/(1-2*vv)*vv
dsde(2,2)=EY/(1+vv)/(1-2*vv)*(1-vv)
dsde(2,3)=EY/(1+vv)/(1-2*vv)*vv
dsde(3,1)=EY/(1+vv)/(1-2*vv)*vv
dsde(3,2)=EY/(1+vv)/(1-2*vv)*vv
dsde(3,3)=EY/(1+vv)/(1-2*vv)*(1-vv)
dsde(4,4)=EY/(1+vv)
dsde(5,5)=EY/(1+vv)
dsde(6,6)=EY/(1+vv)
RETURN
END

SUBROUTINE CSS(tLam,tmuA,Ee,SS)
External Ltensor, T4ddT2
DOUBLE PRECISION tL(3,3,3,3),Ee(3,3)
DOUBLE PRECISION SS(3,3),tLam,tmuA
CALL Ltensor(tLam,tmuA,tL)
CALL T4ddT2(SS,tL,Ee,1.0d0)
```

```
RETURN
END
SUBROUTINE Ltensor(tLam,tmuA,tL)
External delta
DOUBLE PRECISION tL(3,3,3,3),tLam,tmuA,delta
INTEGER i,j,k,l
do i=1,3
do j=1,3
do k=1,3
do l=1,3
tL(i,j,k,l)=tLam*delta(i,j)*delta(k,l)
& +tmuA*(delta(i,k)*delta(j,l)+delta(i,l)*delta(j,k))
end do
end do
end do
end do
RETURN
END
SUBROUTINE T4ddT2(Tres,T4,T2,fact)
DOUBLE PRECISION T4(3,3,3,3),T2(3,3),Tres(3,3)
DOUBLE PRECISION fact
INTEGER i,j,k,l
do i=1,3
do j=1,3
Tres(i,j)=0.0d0
do k=1,3
do l=1,3
Tres(i,j)=Tres(i,j)+fact*T4(i,j,k,l)*T2(k,l)
end do
end do
end do
end do
RETURN
END
c
SUBROUTINE Rvoight1(Ts,Ttt)
Double Precision Ts(6),Ttt(3,3)
Ttt(1,1)=Ts(1)
Ttt(2,2)=Ts(2)
Ttt(3,3)=Ts(3)
Ttt(1,2)=Ts(4)
Ttt(2,3)=Ts(5)
Ttt(1,3)=Ts(6)
Ttt(2,1)=Ts(4)
Ttt(3,2)=Ts(5)
Ttt(3,1)=Ts(6)
RETURN
END
FUNCTION delta(i,j)
DOUBLE PRECISION delta
```

```
INTEGER i,j
IF (i.eq.j) THEN
delta=1.0d0
ELSE
delta=0.0d0
ENDIF
RETURN
END
```

Appendix 7

Input File of the Uniaxial Test with Strain-Hardening Model in Section 6.2

```
/COM, VALIDATED USERMAT BY UNIAXIAL TEST
/UPF, USERMAT.F
/PREP7
!=====================================================!
! MATERIAL PARAMETERS FOR STRAIN-HARDENING MODEL
!=====================================================!
C1 = 288800 !E
C2 = 0.3 !V
C3 = 5.0E-14 !KS
C4 = 5 !NS

TB, USER, 1, , 4
TBDATA, 1, C1, C2, C3, C4
TB, STATE, 1, , 6

!=====================================================!
! GEOMETRY AND MESH
!=====================================================!
ET, 1, 185
BLOCK, 0, 1, 0, 1, 0, 1
ESIZE, , 1
VMESH, ALL

!=====================================================!
! BOUNDARY CONDITIONS
!=====================================================!
! UNIAXIAL TEST SETUP
NSEL, S, LOC, X
D, ALL, UX
NALL
NSEL, S, LOC, Y
D, ALL, UY
NALL
NSEL, S, LOC, Z
D, ALL, UZ
NALL
FINI
```

```
!=====================================================!
! SOLUTION SETTING
!=====================================================!
/SOLU
NROPT, UNSYM
NSUBST, 30, 100, 20
TIME, 1
NSEL, S, LOC, Y, 1
D, ALL, UY, 0.009
NALL
OUTRES, ALL, ALL
OUTRES, SVAR, ALL
SOLVE

FINI
!=====================================================!
! PRINT OUT RESULTS
!=====================================================!
/POST26
ESOL, 2 , 1, , S, Y, S_Y
ESOL, 8, 1, , EPEL, Y
/OUT,
PRVAR, 8, 2
FINISH
```

Appendix 8

Input File of Curve-Fitting of the Ogden Model in Section 8.3

```
/COM, CURVE-FITTING OF OGDEN MODEL
*CREATE, CFHY-OG-U.EXP ! INPUT THE EXPERIMENTAL DATA
! ENGINEERING STRAIN ENGINEERING STRESS (MPA)
0.265384615  0.2125
0.403846154  0.3125
0.588461538  0.4
0.853846154  0.475
1.153846154  0.6
1.396153846  0.675
1.996153846  0.8625
2.55  1.0375
2.988461538  1.225
3.738461538  1.5625
4.303846154  1.9375
4.707692308  2.3
5.076923077  2.675
5.342307692  3
5.538461538  3.3625
5.792307692  3.7125
6.011538462  4.0625
6.092307692  4.45
6.173076923  4.7875
6.334615385  5.1375
6.403846154  5.5375
*END

*CREATE, CFHY-OG-B.EXP ! INPUT THE EXPERIMENTAL DATA
! ENGINEERING STRAIN ENGINEERING STRESS (MPA)

0.173076923  0.3
0.288461538  0.425
0.438461538  0.5125
0.692307692  0.6375
0.934615385  0.75
1.465384615  0.975
1.996153846  1.225
2.4    1.4375
2.734615385  1.6625
3.034615385  1.925
```

```
3.219230769  2.1875
3.403846154  2.4
*END

/PREP7
! DEFINE UNIAXIAL DATA
TBFT, EADD, 1, UNIA, CFHY-OG-U.EXP
TBFT, EADD, 1, BIAX, CFHY-OG-B.EXP
! DEFINE MATERIAL: TWO PARAMETER OGDEN MODEL
TBFT, FADD, 1, HYPER, OGDEN, 2
! SET THE INITIAL VALUE OF COEFFICIENTS
TBFT, SET, 1, HYPER, OGDE, 2, 1, 1
TBFT, SET, 1, HYPER, OGDE, 2, 2, 1
TBFT, SET, 1, HYPER, OGDE, 2, 3, 1
TBFT, SET, 1, HYPER, OGDE, 2, 4, 1

! DEFINE SOLUTION PARAMETERS IF ANY
TBFT, SOLVE, 1, HYPER, OGDEN, 2 , 1, 500

/OUT
! PRINT THE RESULTS
TBFT, LIST, 1
TBFT, FSET, 1, HYPER, OGDEN, 2
TBLIS, ALL, ALL

/OUT
CFIT, , DFLAG, FITTEDDATLIST
CFIT, DEBUG
FINISH
```

Appendix 9

Input File of the Liver Soft Tissue Model in Section 9.1

```
/COM, SIMULATION OF LIVER VISCOELASTICITY
/COM, UNITS (N, MM, MPA)
/PREP7
K, 1, -3.50, 0.00, 0.00
K, 2, -2.00, 1.50, 0.00
K, 3, -0.5, 2.0, 0.00
K, 4, 1.50, 1.50, 0.00
K, 5, 3.00, 0.00, 0.00
K, 6, -5.00, 0.00, 0.00
K, 7, 5.00, 0.00, 0.00
K, 8, 5.00, -0.50, 0.00
K, 9, -5.00, -0.50, 0.00
K, 10, -0.45151, 3
BSPLIN, 1, 2, 3, 4, 5
L, 1, 5
AL, 1, 2
A, 6, 7, 8, 9
ET, 1, 182
KEYOPT, 1, 3, 2
ESIZE, 0.2
AMESH, ALL
! ********************** MATERIAL CONSTANTS **********
TB, HYPE, 1, 1, 1, Mooney ! Mooney-Rivlin MODEL
TBDATA, 1, 21.76e-3*0.542,-21.76e-3*0.542/2
TB, PRONY, 1, 1, 3, SHEAR ! PRONY'S SERIES
TBDATA, 1, 33.92/117.96, 0.58, 20.16/117.96, 11.07,
26.61/117.96, 140.48

MP, EX, 5, 1E6
MP, NUXY, 5, 0.2
MP, MU, 2, 0.3 ! CONTACT
MAT, 2
R, 3
REAL, 3
ET, 2, 169
ET, 3, 172
KEYOPT, 3, 9, 0
KEYOPT, 3, 10, 0
R, 3,
```

```
RMORE,
RMORE, , 0
RMORE, 0
! GENERATE THE TARGET SURFACE
LSEL, S, , , 3
CM, _TARGET, LINE
TYPE, 2
NSLL, S, 1
ESLN, S, 0
ESURF
CMSEL, S, _ELEMCM
! GENERATE THE CONTACT SURFACE
LSEL, S, , , 2
CM, _CONTACT, LINE
TYPE, 3
NSLL, S, 1
ESLN, S, 0
ESURF
ALLSEL
NSEL, S, LOC, Y, -0.5
D, ALL, UY, 0
NSEL, R, LOC, X, -5
D, ALL, ALL
ALLSEL

CYL4, -0.45151, 2.5, 0.5
ESIZE, 0.1
AMESH, 3

MP, MU, 3, 0
MAT, 3
MP, EMIS, 1, 7.88860905221E-031
R, 5
REAL, 5
ET, 6, 169
ET, 7, 172
R, 5, , , 1.0, 0.1, 0,
RMORE, , , 1.0E20, 0.0, 1.0,
RMORE, 0.0, 0, 1.0, , 1.0, 0.5
RMORE, 0, 1.0, 1.0, 0.0, , 1.0
RMORE, , , , , , 1.0
KEYOPT, 7, 3, 0
KEYOPT, 7, 4, 0
KEYOPT, 7, 5, 3
KEYOPT, 7, 7, 0
KEYOPT, 7, 8, 0
KEYOPT, 7, 9, 1
KEYOPT, 7, 10, 0
KEYOPT, 7, 11, 0
KEYOPT, 7, 12, 0
KEYOPT, 7, 14, 0
```

```
KEYOPT, 7, 18, 0
KEYOPT, 7, 2, 0
! GENERATE THE TARGET SURFACE
LSEL, S, , , 9
LSEL, A, , , 10
CM, _TARGET, LINE
TYPE, 6
NSLL, S, 1
ESLN, S, 0
ESURF
CMSEL, S, _ELEMCM
! GENERATE THE CONTACT SURFACE
LSEL, S, , , 1
CM, _CONTACT, LINE
TYPE, 7
NSLL, S, 1
ESLN, S, 0
ESURF
ALLSEL

ET, 8, 169
ET, 9, 172
KEYOPT, 9, 2, 2
KEYOPT, 9, 12, 5
R, 9

! PILOT NODE
TSHAP, PILOT
N, 9999, -0.45151, 2.5
TYPE, 8
REAL, 9
E, 9999

TYPE, 9
REAL, 9
ASEL, S, AREA, , 3
NSLA, S, 1
ESLN, S
ESURF
ALLSEL, ALL
D, 57, UX
ALLSEL

ASEL, S, AREA, , 2, 3
ESLA, S
EMODIF, ALL, MAT, 5
ALLSEL
FINISH
/SOLU
TIME, 0.01
NLGEOM, ON
```

```
NSUBST, 50, 1000, 30
KBC, 0
D, 9999, UY, -0.5
D, 9999, UX, 0, , , , ROTZ
ALLSEL
OUTRES, ALL, ALL
SOLVE
TIME, 500
SOLVE
FINISH
```

Appendix 10

Input File of the Stress Evolution of Glass Tube in Section 9.2

```
/COM, SIMULATION OF STRESS EVOLUTION OF GLASS TUBE
/COM, UNITS(N, M, PA, S)
/PREP7
H = 0.05
ROUT = 0.06
RIN = 0.03

ET, 1, 212
KEYOPT, 1, 3, 2
KEYOPT, 1, 11, 1

RECTNG, RIN, ROUT, 0, H
ESIZE, 0.005
AMESH, 1

MP, EX, 1, 72.5E9
MP, NUXY, 1, 0.3
MP, DENS, 1, 2500

TB, THERM, 1, , , COND
TBDATA, 1, 1

TB, PRONY, 1, , 3, SHEAR
TBDATA, 1, 0.422, 0.0689, 0.423, 0.0065, 0.155, 0.0001

TB, SHIFT, 1, , 6, FICT
TBDATA, 1, 618, 6.45E4, 0.53
TBDATA, 4, 618, 0.108, 3.0, 618, 0.443, 0.671
TBDATA, 10, 618, 0.166, 0.247, 618, 0.161, 0.091
TBDATA, 16, 618, 0.046, 0.033, 618, 0.076, 0.008

MPTEMP, 1, 270
MPTEMP, 2, 550
MP, CTEX, 1, 6.47E-6
MP, CTEX, 1, 34.3E-6

NSEL, S, LOC, Y
D, ALL, UY, 0
ALLSEL
```

```
*DIM, TINN, TABLE, 5, 1, 1, TIME
TINN(1, 0) = 0
TINN(1, 1) = 620
TINN(2, 0) = 1
TINN(2, 1) = 620
TINN(3, 0) = 10
TINN(3, 1) = 520
TINN(4, 0) = 20
TINN(4, 1) = 25
TINN(5, 0) = 25
TINN(5, 1) = 25

*DIM, TOUT, TABLE, 5, 1, 1, TIME
TOUT(1, 0) = 0
TOUT(1, 1) = 620
TOUT(2, 0) = 1
TOUT(2, 1) = 620
TOUT(3, 0) = 10
TOUT(3, 1) = 420
TOUT(4, 0) = 20
TOUT(4, 1) = 25
TOUT(5, 0) = 25
TOUT(5, 1) = 25

NSEL, S, LOC, X, 30E-3
D, ALL, TEMP, %TINN%
ALLS

NSEL, S, LOC, X, 60E-3
D, ALL, TEMP, %TOUT%
ALLS
TOFFSET, 273
FINISH

/SOLU
TIME, 25
ANTYPE, TRANSIENT
NSUBST, 200, 1000, 80
OUTRES, ALL, ALL
SOLV
FINISH

/POST26
ESOL, 2, 25, , FICT, TEMP
ESOL, 3, 25, , EPTH, X
ESOL, 4, 25, , S, X
ESOL, 5, 25, , S, Y
ESOL, 6, 30, , FICT, TEMP
ESOL, 7, 30, , EPTH, X
```

```
ESOL, 8, 30, , S, X
ESOL, 9, 30, , S, Y
PRVAR, 2, 6
PRVAR, 3, 7
PRVAR, 4, 8
PRVAR, 5, 9
```

Appendix 11

Input File of the Small Punch Test in Section 10.2

```
/COM, SIMULATION OF SMALL PUNCH TEST USING BB MODEL
/COM, UNITS (N, MM, MPA, S)
/PREP7

K, 1
K, 2, 1.25, 1.25
K, 3, 0, 1.25
K, 4, 1.25, 2
LARC, 1, 2, 3, 1.25
L, 2, 4

K, 5, 2, 0.5
K, 6, 2.5, 0
K, 7, 2.5, 0.5
K, 8, 4, 0
LARC, 5, 6, 7, 0.5
L, 6, 8

K, 9, 2, -1
K, 10, 2.5, -0.5
K, 11, 2.5, -1
K, 12, 4, -0.5
LARC, 10, 9, 11, 0.5
L, 12, 10

RECTNG, 0, 4, 0, -0.5

ET, 1, 182

KEYOPT, 1, 3, 1

/COM, DEFINING BB MODEL MATERIAL PROPERTIES
LAMBDAL = 3.25
KAPPA = 6000
MUA = 22.0
MUB = 200
GM = 5.906E-8
M = 20
```

```
TB, BB, 1, , , ISO
TBDATA, 1, MUA, LAMBDAL*LAMBDAL, MUB, LAMBDAL*LAMBDAL
TBDATA, 5, GM, 0, M

TB, BB, 1, , , PVOL
TBDATA, 1, 1/KAPPA

/COM, CREATE MESH
ESIZE, 0.2
MAT, 1
TYPE, 1
REAL, 1
AMESH, 1

/COM, THE 1ST CONTACT PAIRS
ET, 2, 169
ET, 3, 172
KEYOPT, 3, 9, 0
KEYOPT, 3, 10, 1
MP, MU, 2, 0.0

R, 2

MAT, 2
REAL, 2
TYPE, 2

ESIZE, 0.2
LMESH, 1, 2
ALLSEL, ALL

TYPE, 3
LSEL, , , , 9
NSLL, , 1
ESLN, , 0
ESURF
ALLS

/COM, THE 2ND CONTACT PAIRS
ET, 4, 169
ET, 5, 172
KEYOPT, 5, 9, 0
KEYOPT, 5, 10, 1
MP, MU, 4, 0.0
R, 4

MAT, 4
REAL, 4
TYPE, 4

ESIZE, 0.2
```

```
LMESH, 3, 4
ALLSEL, ALL

TYPE, 5
LSEL, , , , 9
NSLL, , 1
ESLN, , 0
ESURF
ALLS

/COM, THE 3RD CONTACT PAIRS
ET, 6, 169
ET, 7, 172
KEYOPT, 7, 9, 0
KEYOPT, 7, 10, 1
MP, MU, 6, 0.0

R, 6

MAT, 6
REAL, 6
TYPE, 6

ESIZE, 0.2
LMESH, 5, 6
ALLSEL, ALL

TYPE, 7
LSEL, , , , 7
NSLL, , 1
ESLN, , 0
ESURF
ALLS

/COM, BOUNDARY CONDITION
/COM, RIGID PUNCH
LSEL, , , , 1, 2
NSLL, , 1
D, ALL, UY, -4
D, ALL, UX, 0.0
ALLS

/COM, RIGID DIE FACE
LSEL, , , , 3, 6
NSLL, , 1
D, ALL, UY, 0.0
D, ALL, UX, 0.0
ALLS

/COM, LEFT SIDE
NSEL, , LOC, X, 0
```

```
NSEL, A, LOC, X, 4
D, ALL, UX, 0.0
ALLS
/COM, CHECK THE CONTACT DEFINITION
CNCHECK
ELIST
FINISH

/SOLUTION

NLGEOM, ON
TIME, 480 ! LOADING SPEED 0.5MM/MIN
NSUBST, 50, 500, 25
OUTRES, ALL, ALL
SOLV
FINISH
```

Appendix 12

Input File of the Rubber Tire Damage Model in Section 11.3

```
/COM, SIMULATION OF TIRE DAMAGE (MULLINS EFFECT)
/COM, UNITS (LBF, INCH, PSI)
/PREP7
R1=2.0 ! EXTERIOR RADIUS OF TIRE, INCH
R2=1.0 ! INTERIOR RADIUS OF TIRE
TK=1.0 ! THICKNESS
L=4.0  ! LENGTH FOR ROAD
PI=3.1416
ET, 1, SOLID185
! MATERIAL PROPERTIES (YEOH)
TB, HYPER, 1, , 3, YEOH    !ACTIVATE 3 TERM YEOH DATA TABLE
TBDATA, 1, 191.322  ! C1
TBDATA, 2, 4.9415   ! C2
TBDATA, 3, 0.5346   !C3
TBDATA, 3, 5.0E-5
TB, CDM, 1, , 3, PSE2 ! MODIFIED OGDEN ROXBURGH MULLINS EFFECT
TBDATA, 1, 2.34, 10.0, 0.222
MP, DENS, 1, 0.043
! CREATE GEOMETRY OF TIRE
CYL4, 0, 0, R2, 0, R1, 180, TK
LESIZE, 5, , , 45
LESIZE, 6, , , 4
LESIZE, 11, , , 2
MSHAPE, 0, 3D
MSHKEY, 1
VMESH, ALL
VSYMM, Y, ALL
ALLSEL
NUMMRG, NODE
ALLSEL
! DEFINE COUPLING OF CENTRAL POINT WITH INTERNAL SURFACE
ET, 2, 170
ET, 3, 174
KEYOPT, 3, 2, 2
KEYOPT, 3, 12, 5
R, 3
! PILOT NODE
TSHAP, PILOT
N, 9999, 0, 0, 0    ! CENTER OF TIRE
```

```
TYPE, 2
REAL, 3
E, 9999
TYPE, 3
REAL, 3
ASEL, S, AREA, , 4, 10, 6
NSLA, S, 1
ESLN, S
ESURF
ALLSEL, ALL
! CREATE ROAD SURFACE
K, 101, -L/2, -R1, -L/2+TK
K, 102, L/2, -R1, -L/2+TK
K, 103, L/2, -R1, L/2
K, 104, -L/2, -R1, L/2
A, 104, 103, 102, 101
ET, 4, 170
ET, 5, 174
R, 4, , , ,
TYPE, 4
REAL, 4
ESIZE, 3
AMESH, 13
ALLSEL, ALL
TYPE, 5
REAL, 4
ASEL, S, AREA, , 3, 9, 6
NSLA, S, 1
ESLN, S
ESURF
ALLSEL
! BOUNDARY CONDITIONS
D, 9999, UX, 0, , , , UY, UZ, ROTX, ROTY
D, 9999, ROTZ, 0
ALLSEL
FINISH
/SOLV
ANTYPE, 0
NLGEOM, ON
TIME, 1
OUTRES, ALL, ALL
NSUBST, 10, 1000, 10
KBC, 0
ASEL, S, AREA, , 13
NSLA, S, 1
D, ALL, UX, 0, , , , UZ, ROTX, ROTY, ROTZ
D, ALL, UY, 0.2
ALLSEL
SOLVE
TIME, 10
OUTRES, ALL, ALL
```

```
NSUBST, 100, 1000, 100
KBC, 0
D, 9999, UX, 0, , , , UY, UZ, ROTX, ROTY
D, 9999, ROTZ, 6*PI
ALLSEL
SOLVE
FINISH
/POST1
NN=110
*DIM, FYY, ARRAY, NN
*DIM, TT, ARRAY, NN
*DIM, MZZ, ARRAY, NN
*DO, I, 1, 10
SET, 1, I
*GET, FYY(I), NODE, 9999, RF, FY
*GET, TT(I), ACTIVE, 0, SET, TIME
*GET, MZZ(I), NODE, 9999, RF, MZ
*ENDDO
*DO, I, 1, 100
II=I+10
SET, 2, I
*GET, FYY(II), NODE, 9999, RF, FY
*GET, TT(II), ACTIVE, 0, SET, TIME
*GET, MZZ(II), NODE, 9999, RF, MZ
*ENDDO
```

Appendix 13

Input File of the Breast Tumor Model in Section 12.2

```
/COM, SIMULATION OF A BREAST WITH A TUMOR
/COM, UNITS (N, M, PA, S)
/UPF, USERHYPER.F
/PREP7
PCIRC, 0.1, , 0, 90,
PCIRC, 0.08, , 0, 90,
RECTNG, 0.0, 0.1, -0.05, 0,
RECTNG, 0.0, 0.08, -0.03, 0,
RECTNG, 0.0, 0.08, -0.05, -0.03
PCIRC, 0.02, , -90, 90,
AOVLAP, ALL

TB, HYPER, 1, , 4, USER
TBDATA, 1, 2.79E3, 1.12E3, 1.11, 1E-7 ! ADIPOSE TISSUE
MP, DENS, 1, 0.75E3

TB, HYPER, 2, , 4, USER
TBDATA, 1, 5.38E3, 2.29E3, 2.58, 1E-7 ! FIBROGLANDULAR TISSUE
MP, DENS, 2, 1E3

TB, HYPER, 3, , 4, USER
TBDATA, 1, 8.20E3, 2.15E3, 3.39, 1E-7 ! TUMOR TISSUE
MP, DENS, 3, 1E3

ET, 1, 182
KEYOPT, 1, 3, 1
KEYOPT, 1, 6, 1
MSHAPE, 0, 2D
ESIZE, 0.005
MAT, 3
AMESH, 7, 8

MAT, 2
AMESH, 10, 12, 2

MAT, 1
AMESH, 9, 13, 2

NSEL, S, LOC, Y, -0.05
```

```
D, ALL, UY, 0
ALLSEL

NSEL, S, LOC, X, 0.1
NSEL, R, LOC, Y, -0.03, 0
SF, ALL, PRES, 1000
ALLSEL
FINISH

/SOLU
NLGEOM, ON
NSUBST, 50, 1000, 30
TIME, 1
OUTRES, ALL, ALL
SOLVE
/POST1
SET, LAST
PLNSOL, U, SUM
```

Appendix 14

Input File of the Soil Excavation in Section 14.3

```
/COM, SIMULATION OF SOIL EXCAVATION
/COM, UNITS (N, M, PA, S)
/PREP7

ET, 1, 182
KEYOPT, 1, 3, 2      ! PLANE STRAIN CONDITIONS

RECTNG, 0, 5, 0, -6

ESIZE, 1
MSHKEY, 1
AMESH, 1
! SET MATERIAL DENSITY
DENS, 1, 1.2

! ELASTIC MATERIAL LAW
KAPPA = 0.054
NU0 = 0.35
E0 = 0.34

A0 = 20       ! INITIAL SIZE OF THE YIELD SURFACE
AH_MIN = 0.01*A0
EM = 2880
K0 = EM/(1-2*NU0)
PT_EL = K0*KAPPA/(1+E0)

!PARAMETER TO VON MISES PLASTICITY
BETA_DRY = 1.0       ! SHAPE OF THE YIELD SURFACE IN DRY PART
[-]
BETA_WET = 1.0       ! SHAPE OF THE YIELD SURFACE IN WETTING
PART [-]
LAMBDA_F = 0.37
KS = 1.0
MC = 1.4

! CREATE AND FILL TBDATA FIELD
TB, SOIL, 1, , , CAMCLAY
TBDATA, 1, LAMBDA_F, MC, A0, AH_MIN,
TBDATA, 5, BETA_DRY, BETA_WET, KS
```

```
!TBDATA, 8, KAPPA, NU0, E0, PT_EL
TB, PELAS, 1, , , POISSON
TBDATA, 1, KAPPA, PT_EL, NU0, E0

NSEL, S, LOC, Y, -6 ! BOTTOM SURFACE FIXED
D, ALL, UY, 0
ALLSEL

NSEL, S, LOC, X, 0  ! ROLLER BOUNDARY
D, ALL, UX
ALLSEL

NSEL, S, LOC, X, 5  ! ROLLER BOUNDARY
D, ALL, UX
ALLSEL

ESEL, S, CENT, X, 0, 3
ESEL, R, CENT, Y, -4.0, 0
CM, EDEAD, ELEM
ALLSEL
FINI

/SOLU
NROPT, FULL
ACEL, 0, 10, 0,      ! SET GRAVITY LOAD
NSUBST, 10, 100, 5
NEQIT, 50     ! SET MAXIMUM NUMBER OF EQUILIBRIUM ITERATIONS
LNSRCH, ON
CNVTOL, F, , 1E-5
TIME, 1
SOLVE

CMSEL, S, EDEAD, ELEM
EKILL, ALL
ALLSEL
OUTRES, ALL, ALL
TIME, 2
NSUBST, 50, 1000, 30
LNSRCH, ON
SOLVE
FINISH

/POST1
SET, LAST
CMSEL, U, EDEAD, ELEM
NSLE, S
PLNSOL, EPPL, EQV
FINISH
```

Appendix 15

Input File of the Tower Subsidence Model in Section 14.4

```
/COM, SIMULATION OF SUBSIDENCE OF A TOWER
/COM, UNITS (N, M, PA)
/PREP7
TB, PELAS, 1, , , POISSON ! POROELASTIC MATERIAL
TBDATA, 1, 0.026, 5.0E3, 0.28, 0.889
TB, SOIL, 1, , , CAMCLAY  ! CAM CLAY MODEL
TBDATA, 1, 0.1174, 1, 103E3, 103E3, 1
TBDATA, 6, 1, 1
MP, DENS, 1, 1923

MP, EX, 2, 2E10
MP, NUXY, 2, 0.3
MP, DENS, 2, 5000

K, 1
K, 2, 30
K, 3, 30, 30
K, 4, 0, 30
K, 5, 0, 20
K, 6, 10, 20
K, 7, 10, 30
K, 8, 8, 30
K, 9, 5, 36
K, 10, 0, 36
K, 11, 0, 30
A, 1, 2, 3, 4
A, 5, 6, 7, 4
A, 8, 9, 10, 11
AOVLAP, 1, 2

ET, 1, 182
KEYOPT, 1, 3, 1

TYPE, 1
MAT, 1
ESIZE, 1
MSHAPE, 0, 2D
AMESH, 2
AMESH, 4
```

```
TYPE, 1
MAT, 2
AMESH, 3
ALLSEL

R, 3
REAL, 3
ET, 2, 169
ET, 3, 172
KEYOPT, 3, 9, 1
KEYOPT, 3, 10, 0
R, 3,
RMORE,
RMORE, , 0
RMORE, 0
! GENERATE THE TARGET SURFACE
LSEL, S, , , 12
CM, _TARGET, LINE
TYPE, 2
NSLL, S, 1
ESLN, S, 0
ESURF
CMSEL, S, _ELEMCM
! GENERATE THE CONTACT SURFACE
LSEL, S, , , 7
CM, _CONTACT, LINE
TYPE, 3
NSLL, S, 1
ESLN, S, 0
ESURF
ALLSEL
! BOUNDARY CONDITIONS
NSEL, S, LOC, X, 0
D, ALL, UX, 0
NSEL, S, LOC, Y, 0
D, ALL, UY, 0
NSEL, S, LOC, X, 30
D, ALL, UX, 0
ALLSEL
FINISH

/SOLU
ACEL, , 9.81 ! SET STANDARD GRAVITY FOR DEAD LOAD
ALLSEL
ANTYPE, STATIC, NEW
NLGEOM, ON
NSUBST, 100, 5000, 50
TIME, 1
OUTRES, ALL, ALL
OUTRES, SVAR, ALL
ALLS
SOLVE
FINI
```

Appendix 16

Input File of the Concrete Slump Test in Section 15.2

```
/COM, SIMULATION OF CONCRETE SLUMP TEST BY DRUCKER-PRAGER
MODEL
/PREP7

! CREATE MATERIAL MODEL
MP, EX, 1, 2.25
MP, NUXY, 1, 0.125
MP, DENS, 1, 0.1

! EDP MDOEL
BETA = 0.581 ! DRUCKER-PRAGER PARAMETER (SLOPE OF CONE)
SIGMA_Y = 1.844E-3  ! DRUCKER-PRAGER PARAMETER (STRENGTH)
TB, EDP, 1, , , LYFUN
TBDATA, 1, BETA, SIGMA_Y

ET, 1, 182
KEYOPT, 1, 3, 1 ! AXISYMMETRICAL

K, 1,
K, 2, 0.10
K, 3, 0.05, 0.3
K, 4, 0, 0.3
A, 1, 2, 3, 4
LSEL, S, LINE, , 2, 4, 2
LESIZE, ALL, , , 10, , , , , 0
LSEL, S, LINE, , 1, 3, 2
LESIZE, ALL, , , 5, , , , , 0
ALLSEL
MSHKEY, 1
AMESH, 1

! SET BOUNDARY CONDITIONS
NSEL, S, LOC, Y, 0
D, ALL, UY
ALLS
NSEL, S, LOC, X, 0
D, ALL, UX
ALLS
NSEL, ALL
```

```
! DEFINE LOADING
ACEL, , 0.666 ! SET STANDARD GRAVITY FOR DEAD LOAD
ALLSEL
FINISH
/SOLU
! SOLVE
NLGEOM, ON
NSUBST, 300, 5000, 50
TIME, 1
OUTRES, ALL, ALL
OUTRES, SVAR, ALL
ALLS
SOLVE
FINI
/POST1
SET, LAST
PLNSOL, EPPL, Y
```

Appendix 17

Input File of the Soil-Arch Interaction Model in Section 15.3

```
/COM, SIMULATION OF SOIL-ARCH
/COM, UNITS (N, MM, MPA)
/PREP7
PCIRC, 2090, 1875, 36.87, 143.13,
K, 5, 1670+1500, 1250, 0
K, 6, -1670-1500, 1250, 0
K, 7, 1670+1500, 3090, 0
K, 8, -1670-1500, 3090, 0
L, 1, 5
L, 5, 7
L, 7, 8
L, 8, 6
L, 6, 2
LSEL, S, LINE, , 5, 9
LSEL, A, LINE, , 1
AL, ALL
ET, 1, 182
KEYOPT, 1, 3, 2
MSHAPE, 0, 2D
ESIZE, 125
AMESH, 2
ADELE, 1
PCIRC, 2090, 1875, 36.87, 143.13,
ET, 2, 182
KEYOPT, 1, 3, 2
ESIZE, 125
TYPE, 2
AMESH, 1
/GSAV, CWZ, GSAV, , TEMP
MP, MU, 1, 0.7
MAT, 1
R, 3
REAL, 3
ET, 3, 169
ET, 4, 172
KEYOPT, 4, 9, 0
KEYOPT, 4, 10, 0
R, 3,
RMORE,
```

```
RMORE, , 0
RMORE, 0
! GENERATE THE TARGET SURFACE
LSEL, S, , , 10
CM, _TARGET, LINE
TYPE, 3
NSLL, S, 1
ESLN, S, 0
ESURF
CMSEL, S, _ELEMCM
! GENERATE THE CONTACT SURFACE
LSEL, S, , , 1
CM, _CONTACT, LINE
TYPE, 4
NSLL, S, 1
ESLN, S, 0
ESURF
ALLSEL

MP, EX, 2, 16000
MP, NUXY, 2, 0.2

MP, EX, 3, 1000
MP, NUXY, 3, 0.2
! SET NONLINEAR PARAMETERS FOR INTACT ROCK
BETA=1.05    ! DRUCKER-PRAGER PARAMETER (SLOPE OF CONE)
SIGMA_Y=0.0224      ! DRUCKER-PRAGER PARAMETER (STRENGTH)
TB, EDP, 3, 1, 2, LYFUN    ! EXTENDED DRUCKER-PRAGER MODEL
TBDATA, 1, BETA, SIGMA_Y

TB, EDP, 3, 1, 2, LFPOT    ! ASSOCIATED PLASTIC FLOW
TBDATA, 1, BETA
ESEL, S, TYPE, , 1
EMODIF, ALL, MAT, 3

ESEL, S, TYPE, , 2
EMODIF, ALL, MAT, 2
ALLSEL
FINISH

/SOLU
LSEL, S, LINE, , 6, 8, 2
NSLL, S, 1
D, ALL, UX, 0
LSEL, S, LINE, , 5, 9, 4
NSLL, S, 1
D, ALL, UY, 0
LSEL, S, LINE, , 11, 13, 2
NSLL, S, 1
D, ALL, ALL
ALLSEL
```

```
NSEL, S, NODE, , 97, 99
SF, ALL, PRES, 0.3
ALLSEL
NSUBST, 40, 1000, 20
OUTRES, ALL, ALL
SOLV
```

Appendix 18

Input File of the Concrete Slump Test in Section 16.3

```
/COM, CONCRETE SLUMP TEST BY MOHR-COULOMB MODEL
/PREP7

! CREATE MATERIAL MODEL
MP, EX, 1, 2.25
MP, NUXY, 1, 0.125
MP, DENS, 1, 0.1

!MOHR-COULOMB MODEL
PHIG = 20     ! FRICTION ANGLE
PHIR = 20     ! RESIDUAL FRICTION ANGLE
CG = 1.1547E-3       ! COHESION
CR = 1.0*CG   ! RESIDUAL COHESION
PSIG = 20     ! DILATANCY ANGLE
SIGT = 1.1547E-4     ! TENSILE STRENGTH
SIGTR = 1.0*SIGT     ! RESIDUAL TENSILE STRENGTH

TB, MC, 1, , , BASE
TBDATA, 1, PHIG, CG, PHIR, PSIG, CR
TB, MC, 1, , , RCUT
TBDATA, 1, SIGT, SIGTR

ET, 1, 182
KEYOPT, 1, 3, 1 !AXISYMMETRICAL

K, 1,
K, 2, 0.10
K, 3, 0.05, 0.3
K, 4, 0, 0.3
A, 1, 2, 3, 4
LSEL, S, LINE, , 2, 4, 2
LESIZE, ALL, , , 10, , , , , 0
LSEL, S, LINE, , 1, 3, 2
LESIZE, ALL, , , 5, , , , , 0
ALLSEL

MSHKEY, 1
AMESH, 1
```

```
! SET BOUNDARY CONDITIONS
NSEL, S, LOC, Y, 0
D, ALL, UY
ALLS
NSEL, S, LOC, X, 0
D, ALL, UX
ALLS
NSEL, ALL

! DEFINE LOADING
ACEL, , 0.666 ! SET STANDARD GRAVITY FOR DEAD LOAD
ALLSEL
FINISH

/SOLU
! SOLVE
NLGEOM, ON
NSUBST, 3000, 10000, 1500
NEQIT, 100
PRED, OFF
TIME, 1
OUTRES, ALL, ALL
OUTRES, SVAR, ALL
ALLS
SOLVE
FINI

/POST1
SET, LAST
PLNSOL, EPPL, Y
```

Appendix 19

Input File of the Slope Stability Model in Section 16.4

```
/COM, SIMULATION OF SLOP STABILITY
/COM, UNITS (N, M, PA)
/PREP7
ET, 1, 182
KEYOPT, 1, 3, 2
KEYOPT, 1, 1, 2
K, 1, 0, 0
K, 2, 15, 0
K, 3, 30, 0
K, 4, 40, 0
K, 5, 40, 10
K, 6, 30, 10
K, 7, 15, 10
K, 8, 15, 20
K, 9, 0, 20
K, 10, 0, 10
A, 1, 2, 7, 10
A, 10, 7, 8, 9
A, 2, 3, 6, 7
A, 3, 4, 5, 6
A, 7, 6, 8
AGLUE, ALL
MSHKEY, 1
ESIZE, 1
AMESH, ALL
ALLSEL
MP, EX, 1, 2.8E10
MP, NUXY, 1, 0.2
MP, DENS, 1, 2500
TB, MC, 1, , , BASE ! MOHR COULOMB MODEL
TBDATA, 1, 15.96, 9.20E3, 0, 15.96, 9.20E3     ! FT=1.63
TB, MC, 1, , , RCUT
TBDATA, 1, 0.5E3, 0.5E3
TB, MC, 1, , , MSOL
TBDATA, 1, 2

ALLSEL
NSEL, S, LOC, X, 0
D, ALL, UX, 0
```

```
NSEL, S, LOC, Y, 0
D, ALL, UY, 0
NSEL, S, LOC, X, 40
D, ALL, UX, 0
ALLSEL
ACEL, , 9.81 ! SET STANDARD GRAVITY FOR DEAD LOAD
ALLSEL
FINISH
/SOLV
ANTYPE, STATIC, NEW
NLGEOM, ON
NSUBST, 100, 5000, 50
TIME, 1
OUTRES, ALL, ALL
OUTRES, SVAR, ALL
ALLS
SOLVE
FINI
```

Appendix 20

Input File of the Tunnel Excavation Model in Section 17.3

```
/COM, SIMULATION OF TUNNEL EXCAVATION
/COM, UNITS (N, M, PA)
/PREP7
RECTNG, -5, 5, -5, 5
CYL4, 0, 0, 0, 0, 0.75, 360
ASBA, 1, 2 , , , KEEP
ET, 1, 182
KEYOPT, 1, 3, 2
ESIZE, 0.25
LESIZE, 5, , , 10
MSHAPE, 0, 2D
AMESH, ALL
ALLSEL

MP, EX, 1, 200E6
MP, NUXY, 1, 0.23
MP, DENS, 1, 1810
TB, JROCK, 1, , , BASE     ! BASE MATERIAL FOR JOINTED ROCK
TBDATA, 1, 21.2, 1E6, 10, 21.2, 1E6
TB, JROCK, 1, , , RCUT
TBDATA, 1, 0.1E6, 0.1E6

TB, JROCK, 1, , , RSC
TBDATA, 1, 0

TB, JROCK, 1, , , FPLANE   ! FAILURE PLANE
TBDATA, 1, 15.8, 0.1E6, 0, 15.8, 0.1E6

TB, JROCK, 1, , , FTCUT
TBDATA, 1, 0.05E6, 0.05E6

TB, JROCK, 1, , , FORIE
TBDATA, 1, , -60

NSEL, S, LOC, X, -5
D, ALL, UX, 0
NSEL, S, LOC, Y, -5
D, ALL, UY, 0
ALLSEL
```

```
NSEL, S, LOC, X, 5
D, ALL, UX, 0
NSEL, S, LOC, Y, 5
D, ALL, UY, 0
ALLSEL

INISTATE, SET, DTYP, STRE
INISTATE, DEFI, ALL, , , , -1E6, -1E6, , 0, 0, 0
ALLSEL
FINISH

FINISH

/SOLV
ANTYPE, STATIC, NEW
TIME, 1
NLGEOM, ON
NSUBST, 1
SOLVE
TIME, 2
ASEL, S, AREA, , 2
ESLA, S
EKILL, ALL   ! EKILL COMMAND
ALLSEL
OUTRES, ALL, ALL
NSUBST, 10, 1000, 10
SOLVE
FINI
```

Appendix 21

Input File of One-Dimensional Terzaghi's Problem in Section 18.3

```
/COM, SIMULATION OF ONE DIMENSIONAL TERZAGHI'S PROBLEM
/com, Units (N, M, PA, S)
/prep7
W = 1  ! RECTANGULAR WIDTH
E = 6000      ! YOUNG'S MODULUS
P = 1

ET, 1, 212
KEYOPT, 1, 3, 2      ! PLANE STRAIN CONDITIONS
KEYOPT, 1, 12, 1     ! PERMEABILITY ON

RECTNG, , W, , -12*W! AREA 6W*9W
ESIZE, W
AMESH, 1

MP, EX, 1, E
MP, NUXY, 1, 0.0
BULK = 0.0

FPX = 4E-6    ! SOLID PERMEABILITY
ONE = 1.0

TB, PM, 1
TBDATA, 1, FPX, ONE

NSEL, S, LOC, Y, -12*W      ! BOTTOM SURFACE FIXED
D, ALL, UY, 0
ALLSEL

NSEL, S, LOC, Y, 0
D, ALL, PRES, 0
ALLSEL

D, ALL, UX    ! 1D PROBLEM
ALLSEL

NSEL, S, LOC, Y, 0
SF, ALL, PRES, P
ALLSEL
```

```
FINI

/SOLU
NROPT, UNSYM
OUTPR, ALL, ALL
OUTRES, ALL, ALL
TIME, 0.1

NSUBST, 200, 1000, 100
KBC, 1        ! LOAD APPLIED INSTANTANEOUSLY
SOLVE
FINI

/POST26
NSOL, 2, NODE(0, -12*W, 0), PRES
PLVAR, 2
```

Appendix 22

Input File of the Settlement Model in Section 18.4

```
/COM, SIMULATION OF SOIL CONSOLIDATION
/COM, UNITS (N, M, PA)
/PREP7
ET, 1, 182
KEYOPT, 1, 3, 2
ET, 2, 212
KEYOPT, 2, 3, 2
KEYOPT, 2, 12, 1

K, 1
K, 2, 5
K, 3, 395
K, 4, 400
K, 5, 395, 50
K, 6, 395, 100
K, 7, 400, 125
K, 8, 395, 125
K, 9, 5, 125
K, 10, 0, 125
K, 11, 5, 75
K, 12, 5, 25
K, 13, 0, 25
K, 14, 0, 75
K, 15, 400, 50
K, 16, 400, 100

A, 1, 2, 12, 13
A, 3, 4, 15, 5
A, 2, 3, 5, 12
A, 5, 6, 11, 12
A, 6, 8, 9, 11

CYL4, 100, 58, 0.2
CYL4, 200, 63, 0.2
CYL4, 300, 68, 0.2

ASBA, 4, 6
ASBA, 9, 7
ASBA, 4, 8
```

```
A, 12, 11, 14, 13
A, 11, 9, 10, 14
A, 15, 16, 6, 5
A, 16, 7, 8, 6

LSEL, S, LINE, , 17, 28
LESIZE, ALL, , , 8, , , , , 1
LSEL, S, LINE, , 10, 12, 2
LESIZE, ALL, , , 80, , , , , 1
LSEL, S, LINE, , 11, 13, 2
LESIZE, ALL, , , 12, , , , , 1
ALLSEL
TYPE, 2
MSHAPE, 0, 2D
MSHKEY, 0
MAT, 2
AMESH, 6

MAT, 1
TYPE, 1
MSHKEY, 1
AMESH, 3
AMESH, 5

ALLSEL
MAT, 3
TYPE, 1
MSHKEY, 1
AMESH, 1, 2
AMESH, 4
AMESH, 7, 9
ALLSEL

MP, EX, 1, 65E9
MP, NUXY, 1, 0.25
MP, DENS, 1, 2400

MP, EX, 2, 20E9
MP, NUXY, 2, 0.3
MP, DENS, 2, 2400

FPX=1.0D-10
ONE=1.0

TB, PM, 2, , , PERM !MATERIAL MODEL FOR POROUS MEDIA
TBDATA, 1, FPX, FPX, FPX

MP, EX, 3, 1E9
MP, NUXY, 3, 0.25
MP, DENS, 3, 2400
```

```
NSEL, S, LOC, X, 400
NSEL, A, LOC, X, 0
D, ALL, UX, 0
ALLSEL

NSEL, S, LOC, Y, 0
D, ALL, UY, 0
ALLSEL

NSEL, S, LOC, Y, 125
SF, ALL, PRES, 100E6
ALLSEL

LSEL, S, LINE, , 17, 28
NSLL, S, 1
SF, ALL, PRES, 50E6
ALLSEL
FINISH
/SOLV
NLGEOM, ON
NSUBST, 100, 5000, 20
TIME, 1
KBC, 0
OUTRES, ALL, ALL
OUTRES, SVAR, ALL
ALLS
SOLVE
TIME, 1*3600*3
KBC, 1
NSUBST, 200, 1000, 50
LSEL, S, LINE, , 17, 28
NSLL, S, 1
D, ALL, PRES, 0
SF, ALL, PRES, 0
ALLSEL
SOLVE
FINI
```

Appendix 23

Input File of the Composite Damage Model in Section 19.3

```
/COM, download composite_damage.db from
/COM, www.feabea.net/models/composite_damage.db
/COM, UNITS (N, MM, MPA)
/PREP7
RESUME, 'composite_damage', 'db'
SECTYPE, 1, SHELL
SECDATA, 1.5, 1, 90
SECDATA, 1.5, 1, 0
SECDATA, 1.5, 1, 90
SECDATA, 1.5, 1, 0
SECDATA, 1.5, 1, 90
SECDATA, 1.5, 1, 0
SECDATA, 1.5, 1, 90
SECDATA, 1.5, 1, 0

SECTYPE, 2, SHELL
SECDATA, 1.5, 1, 0
SECDATA, 1.5, 1, 90
SECDATA, 1.5, 1, 0
SECDATA, 1.5, 1, 90
SECDATA, 1.5, 1, 0
SECDATA, 1.5, 1, 90
SECDATA, 1.5, 1, 0
SECDATA, 1.5, 1, 90
LOCAL, 11, 1, 0, 0, 0, 0, 0, 90
ESEL, S, TYPE, , 3, 5, 2
EMODIF, ALL, ESYS, 11
ALLSEL
CSYS, 0

!TYPICAL MATERIAL PROPERTIE FOR E-GLASS/EPOXY (MPA)
E_X=18.4*6894
E_Y=1.62*6894
E_Z=E_Y

PR_XY=0.1279
PR_YZ=0.1331
PR_XZ=PR_XY
```

```
G_XY=0.951*6894
G_YZ=0.528*6894
G_XZ=G_XY

MP, EX, 1, E_X
MP, EY, 1, E_Y
MP, EZ, 1, E_Z
MP, NUXY, 1, PR_XY
MP, NUYZ, 1, PR_YZ
MP, NUXZ, 1, PR_XZ
MP, GXY, 1, G_XY
MP, GYZ, 1, G_YZ
MP, GXZ, 1, G_XZ
MP, DENS, 1, 1
MP, DENS, 2, 1

TB, FCLI, 1, 1, , STRS ! MATERIAL STRENGTHS
TBDATA, 1, 0.283*6894 !FAILURE STRESS, X DIRECTION TENSION
TBDATA, 2, -0.215*6894 !FAILURE STRESS, X DIRECTION COMPRESSION
TBDATA, 3, 6.96E-3*6894 !FAILURE STRESS, Y DIRECTION TENSION
TBDATA, 4, -2.90E-2*6894 !FAILURE STRESS, Y DIRECTION
COMPRESSION
TBDATA, 5, 6.96E-3*6894 !FAILURE STRESS, Z DIRECTION TENSION !
NOT USED BY HASHIN FC
TBDATA, 6, -2.90E-2*6894 !FAILURE STRESS, Z DIRECTION
COMPRESSION ! NOT USED BY HASHIN FC
TBDATA, 7, 1.15E-2*6894 !FAILURE STRESS, XY DIRECTION TENSION
TBDATA, 8, 7.25E-3*6894 !FAILURE STRESS, YZ DIRECTION TENSION
TBDATA, 9, 1.15E-2*6894 !FAILURE STRESS, XZ DIRECTION TENSION !
NOT USED BY HASHIN FC

TB, DMGI, 1, 1, , FCRT ! FC FOR DAMAGE INITIATION
TBDATA, 1, 2, 2, 2, 2 ! MAX STRESS CRITERIA FOR ALL FOUR
FAILURE MODES

TB, DMGE, 1, 1, , MPDG ! DAMAGE EVOLUTION WITH MPDG METHOD
TBTEMP, 0
TBDATA, 1, 0.6 ! 60% FIBER TENSION DAMAGE (40% ULTIMATE
STRENGTH)
TBDATA, 2, 0.6 ! 60% FIBER COMPRESSION DAMAGE (40% ULTIMATE
STRENGTH)
TBDATA, 3, 0.6 ! 60% MATRIX TENSION DAMAGE (40% ULTIMATE
STRENGTH)
TBDATA, 4, 0.6 ! 60% MATRIX COMPRESSION DAMAGE (40% ULTIMATE
STRENGTH)
MP, EX, 2, 18E4
MP, NUXY, 2, 0.3
MP, EX, 3, 180
MP, NUXY, 3, 0.286

ESEL, S, TYPE, , 1
```

```
EMODIF, ALL, MAT, 2
ALLSEL
ESEL, S, TYPE, , 4
EMODIF, ALL, MAT, 3
ALLSEL

NSEL, S, LOC, Z, 130
SF, ALL, PRES, 240
ALLSEL
NSEL, S, LOC, Z, 0
D, ALL, ALL
ALLSEL
FINISH

/SOLU
ANTYPE, 0
NLGEOM, ON
ERESX, NO
OUTRES, ALL, ALL
NSUB, 50, 10000, 20
TIME, 1
SOLV
FINISH

/POST26
NSOL, 3, 28284, U, Z,
PRVAR, 3, , , , , , , , , ,
```

Appendix 24

Input File of the SLB Model in Section 19.4

```
/COM, SLB PROBLEM TO VALIDATE CRACK GROWTH
/COM, UNITS (N, MM, MPA)
/PREP7
DIS1=-5
N1=100
N2=100
N3=10
DL1=177.8
DL2=177.8-34.3
DH=2.032

ET, 1, 182    !* 2D 4-NODE STRUCTURAL SOLID ELEMENT
KEYOPT, 1, 1, 2      !* ENHANCE STRAIN FORMULATION
KEYOPT, 1, 3, 2      !* PLANE STRAIN
ET, 2, 182
KEYOPT, 2, 1, 2
KEYOPT, 2, 3, 2
ET, 3, 202    ! 2D 4-NODE COHESIVE ZONE ELEMENT
KEYOPT, 3, 2, 2
KEYOPT, 3, 3, 2      ! PLANE STRAIN
MP, EX, 4, 135.3E3  ! ANISOTROPIC MATERIAL PROPERTIES OF
LAMINA
MP, EY, 4, 9.0E3
MP, EZ, 4, 9.0E3
MP, GXY, 4, 4.5E3
MP, GYZ, 4, 4.5E3
MP, GXZ, 4, 3.3E3
MP, PRXY, 4, 0.24
MP, PRXZ, 4, 0.24
MP, PRYZ, 4, 0.46

TB, CGCR, 1, , 3, LINEAR ! LINEAR FRACTURE CRITERION
TBDATA, 1, 0.05, 0.05, 0.05

TBLIST, ALL, ALL

RECTNG, 0, DL2, DH  ! DEFINE AREAS
RECTNG, DL2, DL1, DH
RECTNG, 0, DL2, 0, -DH
```

```
LESIZE, ALL, DH/2
MAT, 4
MSHAPE, 0, 2D
AMESH, 1, 3, 2
AMESH, 2
CLOCAL, 12, , , , , 180
CSYS, 0
NSEL, S, LOC, X, 0, DL2-10
NUMMRG, NODES
ESLN
TYPE, 3
CZMESH, , , 1, Y, 0,        ! GENERATE INTERFACE ELEMENTS
ALLSEL, ALL
ASEL, S, AREA, , 1, 2
NSLA, S, 1
NSEL, R, LOC, X, DL2
NUMMRG, NODES
ALLSEL

NSEL, S, LOC, X, 0  ! APPLY CONSTRAINTS
NSEL, R, LOC, Y, 0
D, ALL, ALL
NSEL, ALL
NSEL, S, LOC, X, DL1
D, ALL, UY
ALLSEL
ESEL, S, ENAME, , 202
CM, CPATH1, ELEM

NSEL, S, LOC, X, DL2-10
NSEL, R, LOC, Y, 0
CM, CRACK1, NODE
ALLS
NSEL, S, LOC, X, DL1/2
NSEL, R, LOC, Y, DH ! APPLY DISPLACEMENT LOADING ON TOP
D, ALL, UY, DIS1
ALLSLE
FINISH

/SOLU
TIME, 1
CINT, NEW, 1
CINT, TYPE, VCCT
CINT, CTNC, CRACK1
CINT, NORM, 12, 2
CINT, NCON, 6! NUMBER OF COUNTOURS
CINT, SYMM, OFF

! CRACK GROWTH SIMULATION SET
CGROW, NEW, 1
CGROW, METHOD, VCCT
```

```
CGROW, CID, 1! CINT ID VCCT CALCULATION
CGROW, CPATH, CPATH1       ! CRACK PATH
CGROW, FCOP, MTAB, 1
CGROW, DTIME, 1E-3
CGROW, DTMIN, 1E-4
CGROW, DTMAX, 2E-2
NSUBST, N1, N2, N3
OUTRES, ALL, ALL
SOLV
FINISH

/POST1
SET, LAST
/GOPR
/OUT
PRCI, 1
PRCI, 1, , CEXT
/NOPR
/OUT, SCRATCH
/POST26
NTOP=NODE(DL1/2, DH, 0)
NSEL, ALL
NSOL, 2, NTOP, U, Y, UY
RFORCE, 3, NTOP, F, Y, FY
PROD, 4, 3, , , RF, , , 20
/TITLE, SLB: REACTION AT TOP NODE VERSES PRESCRIBED
DISPLACEMENT
/AXLAB, X, DISP U (MM)
/AXLAB, Y, REACTION FORCE R (N)
/YRANGE, 0, 60
XVAR, 2
PRVAR, UY, RF

/OUT
PRVAR, 2, 3, 4, 5
```

Appendix 25

Input File of the Spur Gear Model with FGM in Section 20.3

```
/COM, SIMULATION OF A SPUR GEAR
/COM, UNITS (N, MM, MPA)
/PREP7
K, 1,
K, 2, 14.1, -2.57
K, 3, 14.9, 1.57
K, 4, 27.4, 4.86
K, 5, 36.4, 9.85
K, 6, 36.4, 15.14
K, 7, 27.4, 20.43
K, 8, 14.75, 23.43
K, 9, 13.88, 27.71
K, 10, 0.125, 24.71
K, 11, 1.125, 12.14
K, 12, 16.5, 22.29
K, 13, 16.5, 2.71
BSPLIN, 6, 7, 12, 8
BSPLIN, 5, 4, 13, 3
BSPLIN, 10, 11, 1
LSTR, 8, 9
LSTR, 9, 10
LSTR, 3, 2
LSTR, 2, 1
LSTR, 6, 5
AL, 1, 2, 3, 4, 5, 6, 7, 8

ET, 1, 182
KEYOPT, 1, 3, 0

ESIZE, 1
AMESH, 1

TB, ELAS, 1  ! DEFINITE THE MATERIAL PROPERTIES
*DO, I, 1, 11
  E%I%=2.01E5*EXP(0.0097*(I-1)*2) ! ASSIGN YOUNG'S MODULUS FOR
THE SUPPORTING POINTS
  TBFIELD, XCOR,    (I-1)*2+14.9
  TBDATA, 1, E%I% , 0.3
```

```
*ENDDO
TBIN, ALGO

F, 1, FY, -1.1368/2
F, 3, FY, -1.1368/2
ALLSEL
LSEL, S, LINE, , 3, 7, 2
NSLL, S, 1
D, ALL, ALL
ALLSEL
FINISH

/SOLU
NSUBST, 1, 1, 1
SOLV
FINISH
```

Appendix 26

Input File of the Orthodontic Wire Model in Section 21.2

```
/COM, SIMULATION OF ORTHODONTIC WIRE
/COM, UNITS (N, MM, MPA)
/PREP7
ET, 1, 182
ET, 2, 185
K, 1, -0.318, 0.216
K, 2, 0.318, 0.216
K, 3, 0.318, -0.216
K, 4, -0.318, -0.216

K, 5, 0, 0
K, 6, 0, 0, 5.5
K, 7, 0, -2, 5.5
K, 67, 0, -2, 10
L, 5, 6
LARC, 6, 7, 67, 1
K, 8, 0, -2, 3
L, 7, 8
K, 9, 0, -3, 2
K, 89, 0, -3, 10
LARC, 8, 9, 89, 1
K, 10, 0, -5, 2
L, 9, 10
K, 11, 0, -6, 3
K, 101, 0, -6, 10
LARC, 10, 11, 101, 1
K, 12, 0, -6, 9.5
L, 11, 12

A, 1, 2, 3, 4
ESIZE, , 4
AMESH, 1
LSEL, S, LINE, , 1, 7
LESIZE, ALL, 0.2
VDRAG, 1, , , , , , 1, 2, 3, 4, 5, 6
VDRAG, 31, , , , , , 7
ACLEAR, 1
ALLSEL
```

```
MP, EX, 1, 5E4
MP, NUXY, 1, 0.3
!DEFINE SMA MATERIAL PROPERTIES
TB, SMA, 1, , , SUPE
TBDATA, 1, 500, 500, 300, 300, 0.07, 0.136

ALLSEL
NSEL, S, LOC, Z, 0
D, ALL, ALL
ALLSEL
FINISH

/SOLU
NLGEOM, ON
NSUBST, 50, 1000, 30
TIME, 1
D, NODE(0, -6, 9.5), UZ, 7
NALL
OUTRES, ALL, ALL
SOLVE
TIME, 2
D, NODE(0, -6, 9.5), UZ, 0
NALL
OUTRES, ALL, ALL
SOLVE
FINISH
```

Appendix 27

Input File of the Vacuum Tight Shape Memory Flange Model in Section 21.3

```
/COM, SIMULATION OF Vacuum-Tight Shape Memory Flange
/COM, UNITS (N, MM, MPA)
/PREP7
! RADIUS OF THE ROD IS 85MM
RR=84
K, 1, 0, 0
K, 2, 0, 900
K, 3, 85, 840
K, 4, 85, 0
K, 5, RR, 670
K, 6, RR+1, 370
K, 7, RR+25, 0
K, 8, RR+140, 0
K, 9, RR+140, 250
K, 10, RR+25, 250
K, 11, RR+15, 370
K, 12, RR+15, 670
A, 1, 2, 3, 4
A, 5, 6, 7, 8, 9, 10, 11, 12
K, 100, 0, 500
VROTAT, 1, 2, , , , , 1, 100, 90

ET, 1, 185
! DEFINE PROPERTIES FOR MATERIAL 1 (METAL)
MP, EX, 1, 53000
MP, NUXY, 1, 0.2

! DEFINE PROPERTIES FOR MATERIAL 2 (ROD)
MP, EX, 2, 74000
MP, NUXY, 2, 0.2

! DEFINE PROPERTIES FOR MATERIAL 5 (SMA)
MP, EX, 5, 60000
MP, NUXY, 5, 0.36
TB, SMA, 5, , , MEFF
TBDATA, 1, 1000, 223, 50, 2.1, 0.04, 45000, 0.0
```

```
ESIZE, 20
VSEL, S, VOLU, , 1
MAT, 2
VSWEEP, 1

VSEL, S, VOLU, , 2
MAT, 1
VSWEEP, 2
ALLSEL
ASEL, S, AREA, , 7
NSLA, S
ESLN, S
EMODIF, ALL, MAT, 5
ALLSEL
MAT, 3
R, 3, 3, 1
REAL, 3
ET, 2, 170
ET, 3, 174
KEYOPT, 3, 9, 1
KEYOPT, 3, 10, 2
R, 3,
! GENERATE THE TARGET SURFACE
ASEL, S, , , 4
CM, _TARGET, AREA
TYPE, 2
NSLA, S, 1
ESLN, S, 0
ESLL, U
ESEL, U, ENAME, , 188, 189
NSLE, A, CT2
ESURF
CMSEL, S, _ELEMCM
! GENERATE THE CONTACT SURFACE
ASEL, S, , , 7
CM, _CONTACT, AREA
TYPE, 3
NSLA, S, 1
ESLN, S, 0
NSLE, A, CT2
ESURF
ALLSEL

NSEL, S, LOC, X, 0
D, ALL, UX,
NSEL, S, LOC, Z
D, ALL, UZ
ALLSEL

NSEL, S, LOC, Y, 0
D, ALL, UY
ALLSEL
```

```
ASEL, S, AREA, , 10
NSLA, S, 1
D, ALL, ALL
FINISH

/SOLVE
ANTYPE, STATIC
OUTRES, ALL, ALL
NROPT, UNSYM
KBC, 0
DELTIM, 0.001
SOLCONTROL, ON
AUTOTS, ON
TIME, 1
ESEL, S, MAT, , 3
EKILL, ALL    ! EKILL COMMAND
ALLSEL
SFA, 7, , PRES, 22
ALLSEL
BFUNIF, TEMP, 223
SOLVE

TIME, 2
DELTIM, 0.001
ESEL, S, MAT, , 3
EALIVE, ALL   ! EALIVE COMMAND
ALLSEL
SFA, 7, , PRES, 0
ALLSEL
BFUNIF, TEMP, 223

SOLVE

TIME, 3
BFUNIF, TEMP, 300
SOLVE
```

Appendix 28

Input File of the Piezoelectric Microaccelerometer Model in Section 22.4

```
/COM, SIMULATION OF MICROACCELEROMETER
/COM, UNITS (UN, UM, MPA)
/PREP7
/TITLE, THE SENSITIVITY IN THE 1G ACCELERATION
ET, 1, 186   ! SELECT ELEMENT TYPE FOR SI MATERIALS
ET, 2, 226, 1001    ! SELECT ELEMENT TYPE FOR PZT
PIEZOELECTRONIC MATERIALS

!MATERIAL PROPERTIES OF THE MASS AND BEAM
MP, DENS, 1, 2.33E-15
MP, EX, 1, 1.9E+5
MP, NUXY, 1, 0.18
STRENGTH=4.2E+2

! **** PIEZOELECTRIC MATERIALS ****
MP, DENS, 2, 7.55E-15

! DEFINE DIELECTRIC CONSTANT [&] ( PF/UM ) ==> F/M = 1E6 PF/
UM)
MP, PERX, 2, 10.443E-3 ! X DIRECTION PERMITTIVITY
MP, PERY, 2, 10.443E-3 ! Y DIRECTION PERMITTIVITY
MP, PERZ, 2, 6.4605E-3 ! Z DIRECTION PERMITTIVITY

!PIEZOELECTRIC STIFFNESS(ELASTIC) COEFFICIENT MATRIX[C] ( UN/
UM2) ==> N/M2 = 1E-6 UN/UM2)
C11=11.425E+4
C12=5.8294E+4
C13=5.8525E+4
C22=11.425E+4
C23=5.8525E+4
C33=9.8181E+4
C44=2.0747E+4
C55=2.0747E+4
C66=2.6042E+4
TB, ANEL, 2  ! INPUT [C] MATRIX FOR PZT
TBDATA, 1, C11, C12, C13
TBDATA, 7, C22, C23
```

```
TBDATA, 12, C33
TBDATA, 16, C44
TBDATA, 19, C55
TBDATA, 21, C66

D31=-93E-6
D33=223E-6
D15=494E-6
E13=2.3687
E23=7.6006
E33=16.422
E51=10.249
TB, PIEZ, 2 ! DEFINE PIEZOELECTRIC TABLE FOR PZT
TBDATA, 3, E13 ! DEFINE MATRIX CONSTANTS
TBDATA, 6, E23
TBDATA, 9, E33
TBDATA, 14, E51

!GIVEN MASS DEMENSION
LMT=800      ! MASS TOP LENGTH (UM)
HM=300 ! MASS HEIGHT (UM)
LMB=LMT-HM*SQRT(2)  ! MASS BOTTOM LENGTH (UM)
H=5

LE=25 ! LENGTH OF THE ELEMENT
BLE=20

! CREATE MASS VOLUME
K, 1, -LMT/2, -LMT/2, 0
K, 2, LMT/2, -LMT/2, 0
K, 3, LMT/2, LMT/2, 0
K, 4, -LMT/2, LMT/2, 0
K, 5, -LMB/2, -LMB/2, -HM
K, 6, LMB/2, -LMB/2, -HM
K, 7, LMB/2, LMB/2, -HM
K, 8, -LMB/2, LMB/2, -HM
V, 1, 2, 3, 4, 5, 6, 7, 8
TYPE, 1
MAT, 1
LSEL, S, LINE, , 5, 11, 2
LESIZE, ALL, , , 5
LSEL, S, LINE, , 1, 4, 3
LESIZE, ALL, LE, , , , , , , 0
ALLSEL
MSHAPE, 0, 3D
MSHKEY, 1
VMESH, 1

!CREATE 15UM THICKNESS MASS TOP VOLUME
K, 9, -LMT/2, -LMT/2, H    ! KEYPOINTS
K, 10, LMT/2, -LMT/2, H
```

```
K, 11, LMT/2, LMT/2, H
K, 12, -LMT/2, LMT/2, H
V, 1, 2, 3, 4, 9, 10, 11, 12, 13

TYPE, 1
MAT, 1
ESIZE, LE
MSHAPE, 0, 3D
MSHKEY, 1
VMESH, 2
ALLS

!GIVEN BEAM DEMENSION
L=400 ! LENGTH OF BEAM SUSPENSION (UM)
B=200 ! WIDTH OF BEAM SUSPENSION (UM)

!GIVEN PZT THIN FILM DEMENSION
LPZT=3*L/10
HPZT=0.5 ! THICKNESS OF PZT THIN FILM (UM) <--- MODIFY
D1=L/10
D2=L-D1-LPZT

APZT=LPZT*B  !CALCULATE XY AREA OF A PZT
CPZT=(6.4605E-3)*(LPZT*B)/HPZT   ! CALCULATE CAPACITANCE OF A
PZT: C= &*A/H

!CREATE BEAM
K, 13, LMT/2, -B/2, 0
K, 14, (L+LMT/2), -B/2, 0
K, 15, (L+LMT/2), B/2, 0
K, 16, LMT/2, B/2, 0
K, 17, LMT/2, -B/2, H
K, 18, (L+LMT/2), -B/2, H
K, 19, (L+LMT/2), B/2, H
K, 20, LMT/2, B/2, H
V, 13, 14, 15, 16, 17, 18, 19, 20

K, 21, B/2, LMT/2, 0
K, 22, B/2, L+LMT/2, 0
K, 23, -B/2, L+LMT/2, 0
K, 24, -B/2, LMT/2, 0
K, 25, B/2, LMT/2, H
K, 26, B/2, L+LMT/2, H
K, 27, -B/2, L+LMT/2, H
K, 28, -B/2, LMT/2, H
V, 21, 22, 23, 24, 25, 26, 27, 28

K, 29, -LMT/2, -B/2, 0
K, 30, -LMT/2, B/2, 0
K, 31, -(L+LMT/2), B/2, 0
K, 32, -(L+LMT/2), -B/2, 0
```

```
K, 33, -LMT/2, -B/2, H
K, 34, -LMT/2, B/2, H
K, 35, -(L+LMT/2), B/2, H
K, 36, -(L+LMT/2), -B/2, H
V, 29, 30, 31, 32, 33, 34, 35, 36

K, 37, B/2, -LMT/2, 0
K, 38, -B/2, -LMT/2, 0
K, 39, -B/2, -(L+LMT/2), 0
K, 40, B/2, -(L+LMT/2), 0
K, 41, B/2, -LMT/2, H
K, 42, -B/2, -LMT/2, H
K, 43, -B/2, -(L+LMT/2), H
K, 44, B/2, -(L+LMT/2), H
V, 37, 38, 39, 40, 41, 42, 43, 44

TYPE, 1
MAT, 1
ESIZE, BLE
LSEL, S, LINE, , 22
LSEL, A, LINE, , 24
LSEL, A, LINE, , 28
LSEL, A, LINE, , 32
LSEL, A, LINE, , 34
LSEL, A, LINE, , 36
LSEL, A, LINE, , 40
LSEL, A, LINE, , 44
LSEL, A, LINE, , 45
LSEL, A, LINE, , 47
LSEL, A, LINE, , 50
LSEL, A, LINE, , 54
LSEL, A, LINE, , 57
LSEL, A, LINE, , 59
LSEL, A, LINE, , 62
LSEL, A, LINE, , 66
LESIZE, ALL, LE
MSHAPE, 0, 3D
MSHKEY, 1
VMESH, 3, 6, 1

! CREATE PZT THIN FILM VOLUME
K, 50, LMT/2+D1, -B/2, H
K, 51, LMT/2+D1+LPZT, -B/2, H
K, 52, LMT/2+D1+LPZT, B/2, H
K, 53, LMT/2+D1, B/2, H
K, 54, LMT/2+D1, -B/2, H+HPZT
K, 55, LMT/2+D1+LPZT, -B/2, H+HPZT
K, 56, LMT/2+D1+LPZT, B/2, H+HPZT
K, 57, LMT/2+D1, B/2, H+HPZT
V, 50, 51, 52, 53, 54, 55, 56, 57
```

```
K, 58, LMT/2+D2, -B/2, H
K, 59, LMT/2+D2+LPZT, -B/2, H
K, 60, LMT/2+D2+LPZT, B/2, H
K, 61, LMT/2+D2, B/2, H
K, 62, LMT/2+D2, -B/2, H+HPZT
K, 63, LMT/2+D2+LPZT, -B/2, H+HPZT
K, 64, LMT/2+D2+LPZT, B/2, H+HPZT
K, 65, LMT/2+D2, B/2, H+HPZT
V, 58, 59, 60, 61, 62, 63, 64, 65

K, 66, B/2, LMT/2+D1, H
K, 67, B/2, LMT/2+D1+LPZT, H
K, 68, -B/2, LMT/2+D1+LPZT, H
K, 69, -B/2, LMT/2+D1, H
K, 70, B/2, LMT/2+D1, H+HPZT
K, 71, B/2, LMT/2+D1+LPZT, H+HPZT
K, 72, -B/2, LMT/2+D1+LPZT, H+HPZT
K, 73, -B/2, LMT/2+D1, H+HPZT
V, 66, 67, 68, 69, 70, 71, 72, 73

K, 74, B/2, LMT/2+D2, H
K, 75, B/2, LMT/2+D2+LPZT, H
K, 76, -B/2, LMT/2+D2+LPZT, H
K, 77, -B/2, LMT/2+D2, H
K, 78, B/2, LMT/2+D2, H+HPZT
K, 79, B/2, LMT/2+D2+LPZT, H+HPZT
K, 80, -B/2, LMT/2+D2+LPZT, H+HPZT
K, 81, -B/2, LMT/2+D2, H+HPZT
V, 74, 75, 76, 77, 78, 79, 80, 81

K, 82, -(LMT/2+D1), -B/2, H
K, 83, -(LMT/2+D1), B/2, H
K, 84, -(LMT/2+D1+LPZT), B/2, H
K, 85, -(LMT/2+D1+LPZT), -B/2, H
K, 86, -(LMT/2+D1), -B/2, H+HPZT
K, 87, -(LMT/2+D1), B/2, H+HPZT
K, 88, -(LMT/2+D1+LPZT), B/2, H+HPZT
K, 89, -(LMT/2+D1+LPZT), -B/2, H+HPZT
V, 82, 83, 84, 85, 86, 87, 88, 89

K, 90, -(LMT/2+D2), -B/2, H
K, 91, -(LMT/2+D2), B/2, H
K, 92, -(LMT/2+D2+LPZT), B/2, H
K, 93, -(LMT/2+D2+LPZT), -B/2, H
K, 94, -(LMT/2+D2), -B/2, H+HPZT
K, 95, -(LMT/2+D2), B/2, H+HPZT
K, 96, -(LMT/2+D2+LPZT), B/2, H+HPZT
K, 97, -(LMT/2+D2+LPZT), -B/2, H+HPZT
V, 90, 91, 92, 93, 94, 95, 96, 97
```

```
K, 98, B/2, -(LMT/2+D1), H
K, 99, -B/2, -(LMT/2+D1), H
K, 100, -B/2, -(LMT/2+D1+LPZT), H
K, 101, B/2, -(LMT/2+D1+LPZT), H
K, 102, B/2, -(LMT/2+D1), H+HPZT
K, 103, -B/2, -(LMT/2+D1), H+HPZT
K, 104, -B/2, -(LMT/2+D1+LPZT), H+HPZT
K, 105, B/2, -(LMT/2+D1+LPZT), H+HPZT
V, 98, 99, 100, 101, 102, 103, 104, 105

K, 106, B/2, -(LMT/2+D2), H
K, 107, -B/2, -(LMT/2+D2), H
K, 108, -B/2, -(LMT/2+D2+LPZT), H
K, 109, B/2, -(LMT/2+D2+LPZT), H
K, 110, B/2, -(LMT/2+D2), H+HPZT
K, 111, -B/2, -(LMT/2+D2), H+HPZT
K, 112, -B/2, -(LMT/2+D2+LPZT), H+HPZT
K, 113, B/2, -(LMT/2+D2+LPZT), H+HPZT
V, 106, 107, 108, 109, 110, 111, 112, 113
!
TYPE, 2
MAT, 2
ESIZE, BLE
VSEL, S, VOLU, , 7, 14
ASLV, S
LSLA, S
LSEL, R, LENGTH, , B
LESIZE, ALL, LE
LSEL, S, LENGTH, , HPZT
LESIZE, ALL, HPZT/2
MSHAPE, 0, 3D
MSHKEY, 1
VMESH, 7, 14, 1

NUMMRG, NODE, 1.0E-4 !MERGES COINCIDENT OR EQUIVALENTLY
DEFINED ITEMS
NUMCMP, NODE !COMPRESSES THE NUMBERING OF DEFINED ITEMS
NUMCMP, ELEM !COMPRESSES THE NUMBERING OF DEFINED ITEMS
NSEL, S, LOC, X, LMT/2+L
CM, D_NODE01, NODE
D, ALL, UX, 0
D, ALL, UY, 0
D, ALL, UZ, 0

NSEL, S, LOC, Y, LMT/2+L
CM, D_NODE02, NODE
D, ALL, UX, 0
D, ALL, UY, 0
D, ALL, UZ, 0
```

```
NSEL, S, LOC, X, -(LMT/2+L)
CM, D_NODE03, NODE
D, ALL, UX, 0
D, ALL, UY, 0
D, ALL, UZ, 0

NSEL, S, LOC, Y, -(LMT/2+L)
CM, D_NODE04, NODE
D, ALL, UX, 0
D, ALL, UY, 0
D, ALL, UZ, 0
ALLSEL

CLOCAL, 12, 0, , , , 90
ESYS, 12
VPLOT, 9, 10
VPLOT
VPLOT, 13, 14
VSEL, S, VOLU, , 9, 10
VSEL, A, VOLU, , 13, 14
VPLOT
ESLV, S
EMODIF, ALL, ESYS, 12
ALLSEL
FINISH

/SOLU
ANTYPE, 0
ALOAD=-9.81E+6
ESEL, S, TYPE, , 2
NSLE, S
NSEL, R, LOC, Z, 5
D, ALL, VOLT, 0
ALLSEL
OUTRES, ALL, ALL,
TIME, 1
NSUBST, 2, 100, 2
KBC, 0

ALLSEL
ACEL, , , ALOAD
SOLVE
FINISH

/POST1
SET, LAST
V11=VOLT(NODE(-760, -100, H+HPZT))+VOLT(NODE(-760, -75,
H+HPZT))+VOLT(NODE(-760, -50, H+HPZT))+VOLT(NODE(-760, -25,
H+HPZT))+VOLT(NODE(-760, 0, H+HPZT))
```

```
V12=VOLT(NODE(-760, 100, H+HPZT))+VOLT(NODE(-760, 75,
H+HPZT))+VOLT(NODE(-760, 50, H+HPZT))+VOLT(NODE(-760, 25,
H+HPZT))
V21=VOLT(NODE(-640, -100, H+HPZT))+VOLT(NODE(-640, -75,
H+HPZT))+VOLT(NODE(-640, -50, H+HPZT))+VOLT(NODE(-640, -25,
H+HPZT))+VOLT(NODE(-640, 0, H+HPZT))
V22=VOLT(NODE(-640, 100, H+HPZT))+VOLT(NODE(-640, 75,
H+HPZT))+VOLT(NODE(-640, 50, H+HPZT))+VOLT(NODE(-640, 25,
H+HPZT))
DV1=(V11+V12-V21-V22)/5
V31=VOLT(NODE(-560, -100, H+HPZT))+VOLT(NODE(-560, -75,
H+HPZT))+VOLT(NODE(-560, -50, H+HPZT))+VOLT(NODE(-560, -25,
H+HPZT))+VOLT(NODE(-560, 0, H+HPZT))
V32=VOLT(NODE(-560, 100, H+HPZT))+VOLT(NODE(-560, 75,
H+HPZT))+VOLT(NODE(-560, 50, H+HPZT))+VOLT(NODE(-560, 25,
H+HPZT))
V41=VOLT(NODE(-440, -100, H+HPZT))+VOLT(NODE(-440, -75,
H+HPZT))+VOLT(NODE(-440, -50, H+HPZT))+VOLT(NODE(-440, -25,
H+HPZT))+VOLT(NODE(-440, 0, H+HPZT))
V42=VOLT(NODE(-440, 100, H+HPZT))+VOLT(NODE(-440, 75,
H+HPZT))+VOLT(NODE(-440, 50, H+HPZT))+VOLT(NODE(-440, 25,
H+HPZT))
DV2=(V31+V32-V41-V42)/5
DV1_2=DV1+DV2

V_OUTPUT=ABS(DV1_2)*4
/OUT,
*STATUS, V_OUTPUT
```

Appendix 29

Input File of the Contact Model in Section 23.2

```
/COM, download nano2.cdb from
/COM, www.feabea.net/models/nano2.cdb
/COM, UNITS (N, MM, MPA)
*DO, I, 1, 20

*IF, I, EQ, 1, THEN
CDREAD, DB, 'nano2', 'cdb', , '', ''
YM0=20E3
*ELSE
/CLEAR,
PARRES,
/PREP7
CDREAD, DB, 'NANO2', 'CDB',    '', ''
R5=3.9427*18.75-21.48
R6=3.9427*22.5-21.48
R7=3.9427*26.25-21.48
R8=3.9427*30-21.48
R5678=(R5+R6+R7+R8)/4*1E-6
YM=R5678/RFTOL*YM0
YM0=YM
MP, EX, I+10, YM
MP, NUXY, I+10, 0.3
ESEL, S, TYPE, , 7
EMODIF, ALL, MAT, I+10
ALLSEL
*ENDIF
FINISH

/SOLU
TIME, 1
NLGEOM, ON
NSUBST, 8, 8, 8
OUTRES, ALL, ALL
SOLV
FINISH

/POST1
RFTOL=0.0
*DO, K, 5, 8
```

```
SET, 1, K
NSEL, S, LOC, Y, 125E-6
*GET, NUM_N, NODE, 0, COUNT       ! GET NUMBER OF NODES
*GET, N_MIN, NODE, 0, NUM, MIN    ! GET MIN NODE NUMBER
RFTOL%K%=0.0
*DO, J, 1, NUM_N, 1 ! OUTPUT TO ASCII BY LOOPING OVER NODES
CURR_N=N_MIN
*GET, RFN, NODE, CURR_N, RF, FY
RFTOL%K%=RFTOL%K%+RFN
*GET, N_MIN, NODE, CURR_N, NXTH
*ENDDO
RFTOL=RFTOL+ABS(RFTOL%K%)
*ENDDO
*STATUS, RFTOL
*STATUS, YM
*STATUS, I
R5=3.9427*18.75-21.48
R6=3.9427*22.5-21.48
R7=3.9427*26.25-21.48
R8=3.9427*30-21.48
R5678=(R5+R6+R7+R8)/4*1E-6
*IF, RFTOL, LT, R5678*1.05, AND, RFTOL, GT, R5678*0.95, THEN
*STATUS, YM0
*EXIT
*ENDIF
PARSAV, SCALAR
*ENDDO
```

Index

For Product Safety Concerns and Information please contact our EU representative GPSR@taylorandfrancis.com
Taylor & Francis Verlag GmbH, Kaufingerstraße 24, 80331 München, Germany

www.ingramcontent.com/pod-product-compliance
Ingram Content Group UK Ltd.
Pitfield, Milton Keynes, MK11 3LW, UK
UKHW022217190926
475702UK00003B/414
9781032566023